The Biology of Fungi

Other Related Books

978-81-7233-519-9	Ascomycetes of Peninsular India	*Pandey, Alaka*
81-7233-453-2	Disease Problems in Vegetable Production	*Gupta, S.K.*
978-81-7233-517-5	Diseases Management in Arid Land Crops	*Lodha, Satish*
978-81-7233-543-4	Diseases of Horticultural Crops: Nematode Problems & their Management	*Reddy, Parvatha,*
978-81-7233-473-4	Essentials of Virology	*Reddy, S. Ram*
978-81-7233-629-5	Fungal Diseases and their Management in Horticultural Crops	*Parvatha Reddy*
81-7233-338-2	Fungi in Human and Animal Health	*Kushwaha, R.K.S*
81-7233-403-6	Fungi: Diversity and Biotechnology	*Rai, M.K.*
81-7233-301-3	Genetic Improvement of Field Crops	*Singh, C.B.*
81-7233-417-6	Indian Erysiphaceae	*Paul, Y.S.*
978-81-7233-525-0	Indian Sarcoscyphaceous	*Pant, D.C.*
81-7233-316-1	Indian Septoriae	*Paul, Y.S.*
978-81-7233-789-6	Integrated Disease Management & Plant Health	*Gupta, V.K.*
978-81-7233-400-0	Integrated Plant Disease Management	*Sharma, R.C.*
978-81-7233-560-1	Integrated Plant Pathology	*S.M. Reddy*
978-81-7233-718-6	Laboratory Exercises in Plant Pathology : An Instructional Kit (2 Vol. Set)	*Baudoin, ABAM*
978-81-7233-706-3	Laboratory Manual of Microbiology	*Kumar, V.*
978-81-7233-766-7	Laboratory Manual of Microbiology, Biochemistry and Molecular Biology	*Saxena, Jyoti*
81-7233-418-4	Mushroom Cultivation Technology	*Gogoi, Robin*
81-7233-410-9	Mycorrhiza	*Prakash, Anil*
978-81-7233-671-4	Mycotoxins Problem and its Management	*Reddy, S.M.*
978-81-7233-792-6	Participatory Technology Development	*Israel Thomas, M*
978-81-7233-701-8	Phytopathogenic Procaryotes and Plant Diseases	*Thind, B.S.*
978-81-7233-518-2	Plant Pathology at a Glance (Encyclopedia of Plant Pathology)	*Tripathi, D.P.*
978-81-7233-586-1	Postharvest Technology of Kinnow	*Sharma, A.D.*
978-81-7233-636-3	Progress in Mycology	*Rai, M.K.*
81-7233-468-0	Seed Borne Diseases - Ecofriendly Management	*Arya, Arun*
81-7233-266-1	Seed Borne Diseases Objectionable in Seed Production and their Management	*Bhale, M.S.*
81-7233-285-8	The Influence of One Plant on Another: Allelopathy	*Molisch, H.*

for more details visit www.scientificpub.com

The Biology of Fungi

Sixth edition

C.T. INGOLD

**Former Professor of Botany,
Birkbeck College, University of London, UK**

H.J. HUDSON

**Lecturer in Plant Sciences,
Cambridge University, UK**

SCIENTIFIC PUBLISHERS (INDIA)

Reprint in India, 2013

Published by:

Scientific Publishers (India)
5 A, New Pali Road, P.O. Box 91
Jodhpur 342 001 (India)

Scientific Publishers (India)
4806/24, Ansari Road, Daryaganj
New Delhi - 110 002 (India)

E-mail: info@scientificpub.com
Website: www.scientificpub.com

ISBN: 978-81-7233-820-6

First Edition, 1961
Sixth Edition 1993

Printed in India

Contents

A colour plate section appears between pages 16 and 17

Preface to the sixth edition

This book has passed through a number of editions each involving some modifications while retaining the general character of the first. The most substantial changes occurred in the fifth (1983), but that is now out-of-date. Mycology has not been immune from the rapid evolution of biology in the past decade and the preparation was a long way beyond the capacity of the older writer (C.T.I.). So a younger one (H.J.H.) has collaborated with him. Substantial changes have been made throughout, particularly in relation to taxonomy but in general Chapters 1–7 have undergone relatively little revision. However, this is not true of Chapters 8–12 where the influence of the younger author is paramount.

The basic idea of the book remains, namely to present fungi as living organisms profoundly affecting the lives of other kinds of living organisms, especially Man. Their study is of the greatest importance and what is more it is great fun.

Many of the diagrams have been retained, but some have been redrawn and new ones added. We are pleased to acknowledge the patience and artistic skills of Mrs Lindsay J. Wilson in producing these. Some coloured plates have also been introduced from photographs taken by H.J.H.

C.T. Ingold and H.J. Hudson
1993

Preface to the fifth edition

In this book I have tried to set out simply my own attitude to mycology and to communicate my belief that fungi are best studied as living organisms in which structure should be considered in relation to function. The mycological taxonomist, the geneticist, the fungal biochemist and the electron-microscopist will continue to make invaluable contributions to mycology, but it will be a sad day when students of fungi no longer marvel at their beauty of structure or cease to enquire into how fungi function as complete living mechanisms.

The original manuscript was written on the Atlantic from Southampton to Montreal and back aboard R.M.S. Ivernia in the summer of 1959. Any merit the book may have owes much to a peaceful and sunny ocean and to the delightful company I enjoyed on board.

In subsequent editions local modifications of a greater or lesser extent have been made. Now, while retaining the basic form, the book has been completely re-written, an operation of very much greater length than the preparation of the original manuscript. A number of the diagrams have been modified and about 30 new ones added. There are now 114 compared with 61 in the first edition. I have especially striven to maintain a close integration of text and illustrations.

Most of the new diagrams are my own, but some of these have already appeared elsewhere. In this connection I have to thank Edward Arnold for permission to reproduce certain illustrations (Figs. 21, 26, 66 and 67) from their 'Studies in Biology', Nos. 88 and 113.

I am also indebted to the Cambridge University Press for allowing me to use, as Fig. 79, a drawing by J. Webster from his *Introduction to Mycology*, and to Boubée et Cie for permission to produce as my final figure an illustration from *Champignons toxiques et hallucinogènes* by the great French mycologist Roger Heim.

C.T. Ingold
1983

Introduction

Traditionally, living organisms have been divided into two kingdoms: Plant and Animal. However, many biologists now recognize five: Prokaryotae, Protoctista, Fungi, Plantae and Animalia. A fundamental difference separates the Prokaryotae, the bacteria and their allies, from the eukaryotic organisms in the other four kingdoms. In prokaryotic types the fine structure of the cell, as revealed by the electron microscope, is quite different from that of animals, green plants, fungi and protoctists. Thus in the bacteria there is no true, membrane-bound nucleus or chromosomes, genetic material consisting of a circular strand of DNA lying in the cytoplasm. Mitochondria and endoplasmic reticulum are missing. The cell surface membrane may be infolded to form mesosomes. Ribosomes are smaller. Flagella, if they occur, are simpler in structure, being made up of individual sub-units of a protein called flagellin. They resemble just one of the 9 + 2 microtubules of eukaryotic flagella. The striking difference between cellular organization in fungi and bacteria can be illustrated by comparing a unicellular fungus, such as the yeast *Saccharomyces* with a bacterium (Fig. 1.1).

The kingdom Protoctista is the least tenable of the five, containing as it does organisms claimed by mycologists, botanists and zoologists. In the Mastigomycotina, the Oomycetes are a case in point. Many mycologists tend to think of them as occupying a rather isolated position in the fungi because of their unusual features. They are retained here because of their similarities with the fungi. Indeed, mycology without Oomycetes would be almost inconceivable.

The term 'fungi' has in the past been used in the wide sense to include the 'slime moulds' and in narrower sense to exclude them. The more restricted use has been adopted in this book. Their taxonomy has always been contested. They are now usually placed in the Myxomycota and Acrasiomycota in the Protoctista. It is true that mycologists have been responsible for most of the research on slime moulds, but they have really no claim to be fungi. Although heterotrophic, they are amoeboid and plasmodial. They ingest food by phagocytosis. They only resemble the fungi in their spore-bearing stages.

Fungi differ fundamentally from green plants in their lack of chlorophyll. They cannot, therefore, photosynthesize their organic food from carbon dioxide and water and are, like animals, forced to live either as saprotrophs on

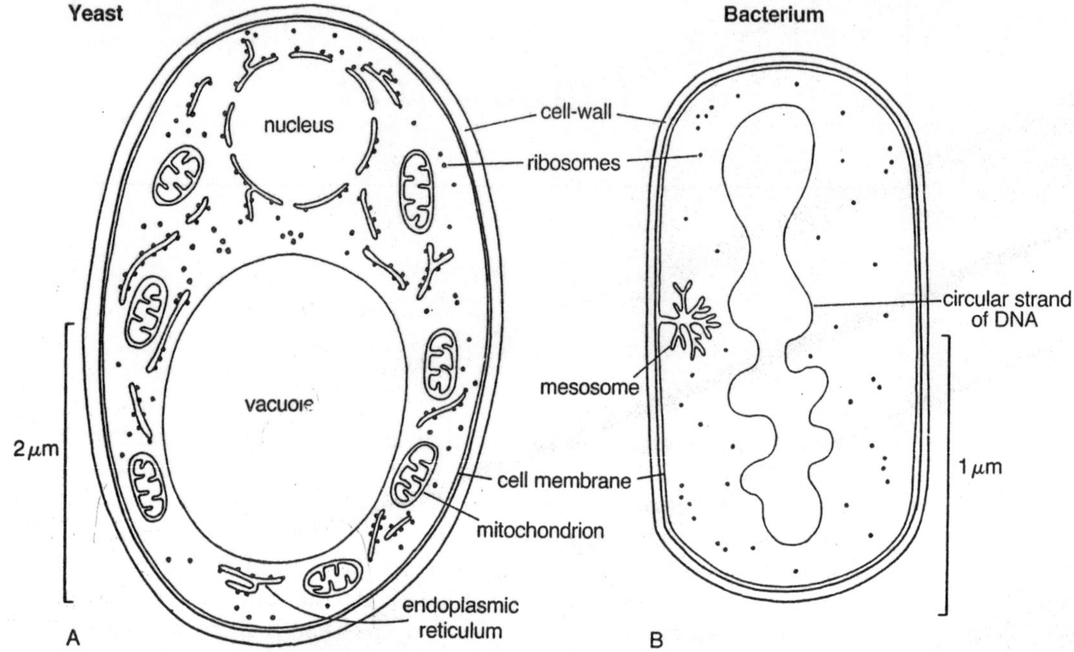

Fig. 1.1. Realistic diagrams indicating the differences in basic cell structure between a yeast (*Saccharomyces*) cell (eukaryotic) and a bacterial cell (prokaryotic).

dead organic matter or as necrotrophic or biotrophic parasites on living organisms or in mutualistic associations with others. Thus, although fungi agree with green plants in having their protoplasm encased in cell walls, they differ essentially in their mode of nutrition.

Species determination in fungi is to a large extent subjective and often controversial, so that any numbers quoted are to be treated with caution. Further, many remain to be discovered. A recent estimate is 64 200 species. In addition the lichens should be included, for these remarkable dual organisms are intimate associations between certain fungi and a simple alga or a cyanobacterium. Some 13 500 fungi occur as lichens.

The large fungal kingdom includes the numerous moulds found on decaying vegetation and the unicellular yeasts abundant on the sugary surface of ripe fruit. In addition there are the water-moulds often seen on dead, floating fish. To the fungi also belong mildews, smuts, rusts and many other plant pathogens. Further, there are the larger fungi: toadstools, bracket polypores, puff-balls and stink-horns so common in woods in autumn and, in addition, the lichens.

In essence a fungus consists of a feeding system of colourless threads (the mycelium), which finally produces spores on specialized reproductive structures. A single branch of the mycelium is known as a hypha which is bounded by a firm hyphal wall.

In studying the form and development of fungi, the mycologist commonly grows them on sterile nutrient agar in petri dishes. A 2% suspension of agar in water, having been sterilized at 120°C, remains fluid down to 38°C when it can be poured into sterile petri dishes. It then sets to a firm jelly. Agar, derived from certain red seaweeds, is a complex carbohydrate which fungi cannot hydrolyse. Appropriate food substances can be incorporated and these are extracted by the fungus without affecting the physical character of the jelly. Frequently malt extract, at a concentration of 1–2%, is included and this contains all the substances needed for fungal growth.

To introduce the question of the origin and development of the mycelium, it is convenient to consider such a common mould as *Mucor*, species of which are abundant in the soil and on dung. They grow quickly and are easy to handle in the laboratory.

The spore of *Mucor*, like that of most fungi, is microscopic, unicellular and full of protoplasm containing a small food reserve. This is in the form of glycogen. In fungi generally, glycogen and fat are the common reserves. Starch does not occur in fungal cells.

Planted on nutrient agar at around 20°C, a spore swells considerably during the first few hours. Then one or more germ tubes grow out as short hyphae which soon branch to form the incipient mycelium (Fig. 1.2A). After a day or two a colony is produced with a circular margin. Grown on a shallow layer of agar in a normal petri dish, the colony appears two-dimensional as a disk. If, however, *Mucor* is cultivated on a deep layer of agar in a beaker, thus allowing unrestricted downward growth, the colony is actually hemispherical (Fig. 1.2B).

Around the edge of the colony are leading hyphae growing outwards roughly in a radial direction. Each elongates by strictly apical growth, a main axis growing faster than its laterals (Fig. 1.3), although, from time to time, a vigorous lateral itself becomes a leader as the circumference of the colony increases.

In most species of *Mucor* and its allies, the mycelium tends to be without cross-walls, but occasional septa appear at a later stage in development, and in most other fungi the mycelium is divided into cells by cross-walls.

It is this branching, feeding mycelium that is the most characteristic feature of fungi and in the following chapter its nutrition, growth and detailed structure will be considered.

After a period of vegetative growth most fungi produce spores on specialized branches of the mycelium. In *Mucor* the system is simple. Sporangia containing spores are formed at the ends of sporangiophores. The sporangiophore starts as an erect aerial hypha. Finally its apical growth ceases and its tip swells into a spherical sporangium delimited from the sporangiophore by a dome-like cross-wall. Within the sporangium the protoplasm cleaves into a large number of equal parts each of which acquires a wall and becomes a spore (Fig. 1.2C). Eventually the spores are liberated and dispersed.

In *Mucor* the spore-bearing apparatus is very simple, but in some fungi, such as the woody bracket polypores often to be seen on tree-trunks, it is large

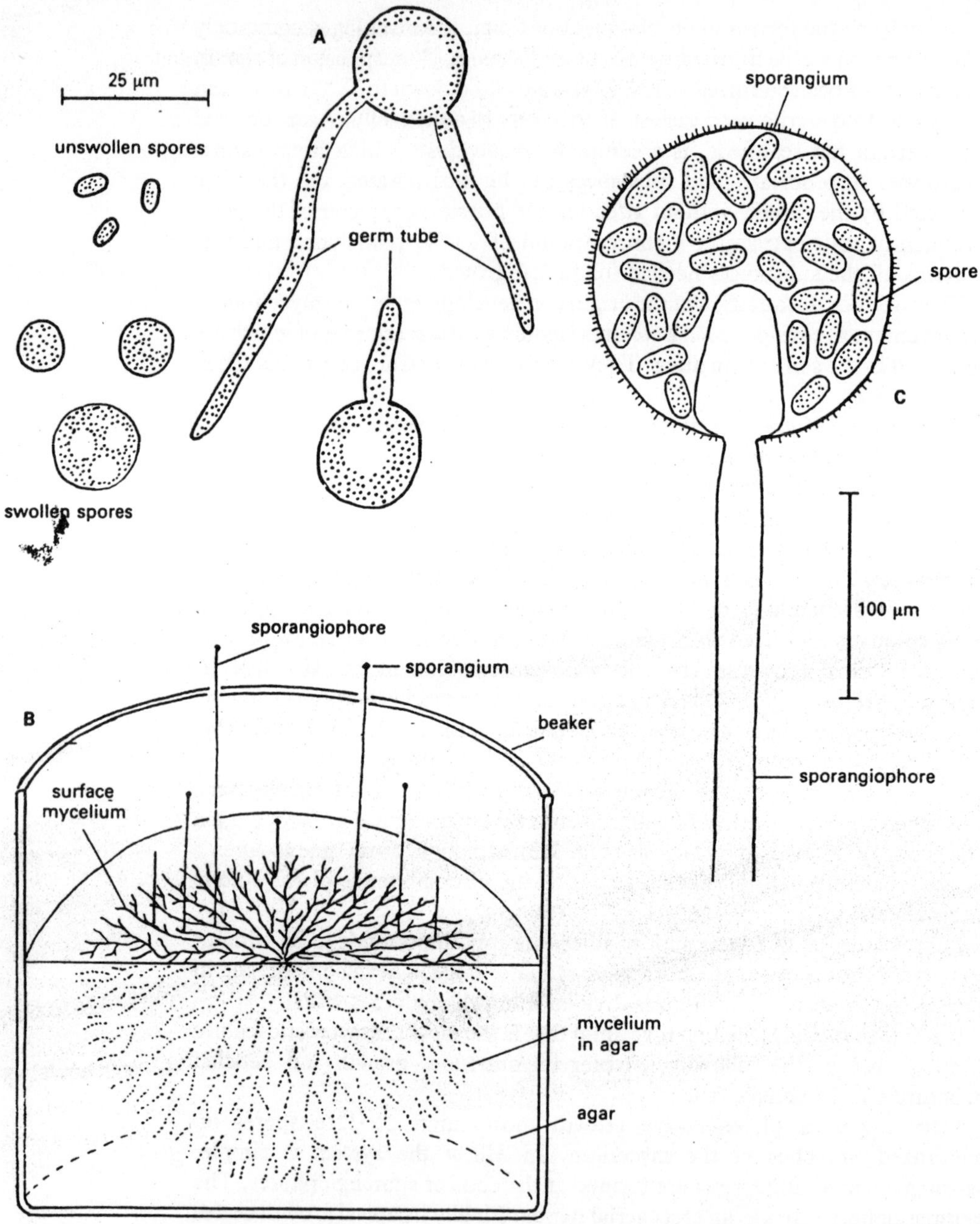

Fig. 1.2. A, *Mucor hiemalis*. Spore germination: spores first swell and then produce one or two germ tubes. B, *M. hiemalis*, diagram of the back half of a beaker half-full of malt agar in which there is a colony of the fungus four days old; submerged mycelium indicated by dotted lines, surface mycelium by solid lines; from the surface mycelium sporangiophores have been produced. C, *M. plasmaticus*, single sporangiophore with spherical sporangium containing spores.

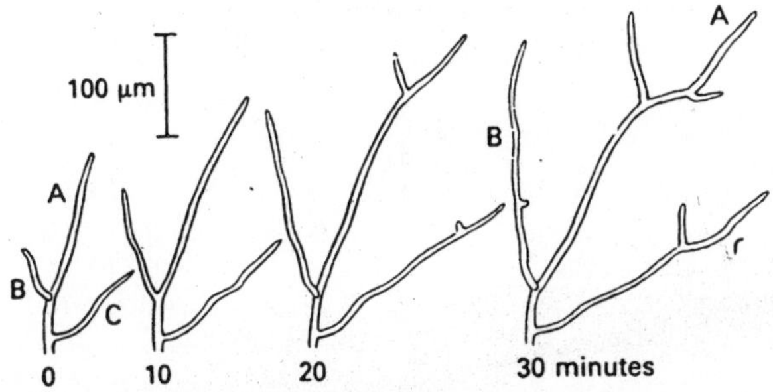

Fig. 1.3. *Mucor hiemalis.* A leading hypha (A) of the mycelium with two laterals (B and C) drawn at 10-minute intervals at 20°C. After 30 minutes A has increased by 120 μm (and produced two laterals), B by 103 μm and C by 90 μm.

and complex. However, these big and elaborate structures are composed entirely of branched, interwoven hyphae. Indeed, the spore-bearing system is always a specialized extension of the mycelium.

In the life of a fungus there are a number of essential phases: spore germination; vegetative growth of the mycelium; development of the spore-bearing apparatus; production of the spores; and finally their liberation and dispersal. Further, like other organisms, the sexual fusion of haploid nuclei and the subsequent meiosis of diploid nuclei are important episodes in the life-cycles of most fungi.

Growth and nutrition

2

Before considering growth and nutrition, it is useful to have a detailed picture of the fungal mycelium. With rare exceptions, growth is strictly apical and, therefore, from the dynamic point of view, it is the hyphal tip that is of most interest. It is the main centre of metabolic activity. Details of the fine structure of the hyphal tip, as revealed by electron-microscopy, vary somewhat in the different groups of fungi. However, the basic organization is the same. Fig. 2.1 shows the appearance in *Pythium*.

The fungal hypha is limited by a hyphal wall (cell-wall). It is a firm aqueous gel with a complex fine structure which, however, has received detailed study only in a few species. It is composed of a range of polymers: glucans (polymers of glucose), chitin (polymer of acetamido-deoxy-glucopyranose) and proteins (polymers of amino-acids). There is an amorphous matrix of glucans filling in a felt or web of long interwoven microfibrils visible only with the EM. The web, which adjoins the plasmalemma, does not extend to the outer surface of the hypha where there is a layer of glucans free from microfibrils. In most fungi the microfibrils are composed of chitin. However, in Mucorales they are of chitosan (chemically rather similar to chitin) and in Oomycetes of a form of cellulose. There is little evidence of cellulose in fungal walls outside this small group. Proteins also occur in the wall especially as glycoprotein, a protein combined with a carbohydrate. In some, such as *Neurospora crassa*, this takes the form of a net, like a coarse stranded but very fine meshed wire netting between the glucan and the chitin. Further, the hyphal wall may have a small amount of lipid.

It is the web of microfibrils that maintains the form of the hypha. One way in which the hyphal wall has been studied is by 'enzyme dissection', a process involving the successive removal of the different components of the wall by the use of the appropriate enzyme. If all but the chitin is digested, the integrity of the hypha is still retained. Some workers have succeeded in digesting the whole of the hyphal wall leaving only the living protoplasts. Under these conditions the protoplasts simply round off.

The hyphal wall is usually regarded as a non-living membrane around the living protoplast, but it is not certain that a sharp distinction can be made between the two. The wall allows diffusion in both directions of water and

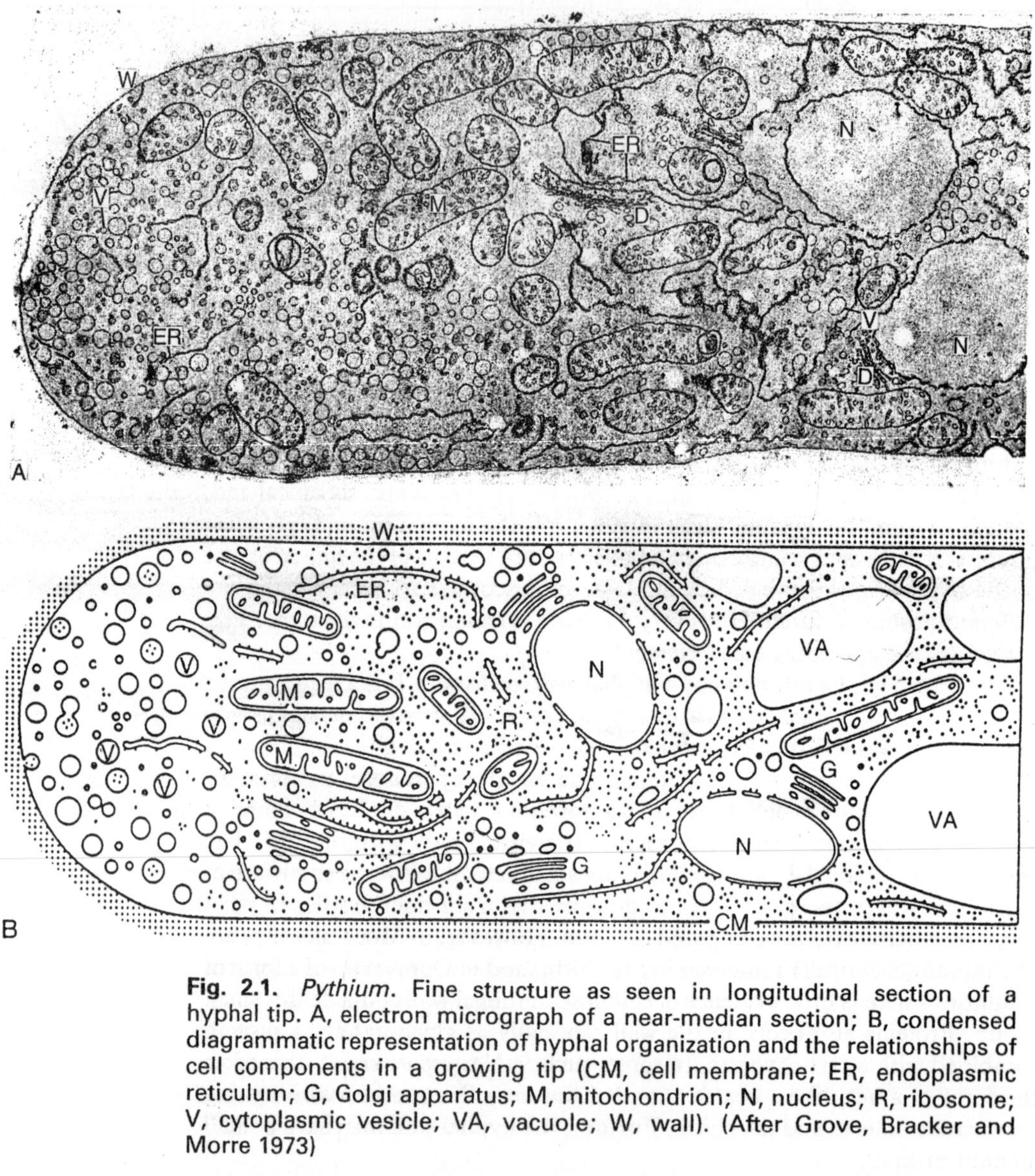

Fig. 2.1. *Pythium.* Fine structure as seen in longitudinal section of a hyphal tip. A, electron micrograph of a near-median section; B, condensed diagrammatic representation of hyphal organization and the relationships of cell components in a growing tip (CM, cell membrane; ER, endoplasmic reticulum; G, Golgi apparatus; M, mitochondrion; N, nucleus; R, ribosome; V, cytoplasmic vesicle; VA, vacuole; W, wall). (After Grove, Bracker and Morre 1973)

dissolved substances. Some of the molecules passing outwards may be quite large. Thus it seems that certain exoenzymes that escape through the wall may have molecular masses around 20 000. Possibly these pass through special pores.

The protoplasm within the wall is delimited by the cell surface membrane. Although the hyphal wall sets limits to what may reach the protoplasm, the cell surface membrane appears to regulate the transport of molecules and ions into and out of the hypha. The two major components of the membrane are lipid and protein. It seems to consist of an ultramicroscopic patchwork of relatively impermeable lipid regions and more permeable proteinaceous ones.

In the lipid areas the membrane appears to be composed of two layers of lipid molecules coated with protein. The membrane is seen under the EM as a single-layered structure and is an example of a 'unit membrane' approximately 8 nm thick.

There may be passive diffusion of certain substances across the membrane, but most transport appears to be active (energy consuming) with the respiratory system of the hypha involved. Active transport seems to be related to carrier molecules which can move across the membrane combining with a solute on one side and liberating it on the other. So far as outward passage of solutes is concerned, bursting vesicles may also effect transport to the outside of the protoplast.

The fine structure of the protoplasm is similar to that in eukaryotic organisms generally. In *Pythium* (Fig. 2.1), as in *Mucor*, cross-walls do not usually occur and there are scattered nuclei. Each of these has a limiting double layer of unit membranes, but there are pores through which, no doubt, messenger RNA molecules, templates of the DNA of the chromosomes, pass with their coded information for synthesis of individual proteins. Outside the nuclei, and sometimes clearly continuous with the nuclear membrane, is abundant endoplasmic reticulum. Attached to this, and also lying free, are numerous ribosomes which, on receipt of messages from the RNA molecules, construct various proteins which are the enzymes vital for the biochemical activity of the hypha. Concentrated at the extreme apex of the hypha are tiny vesicles of uniform size each delimited by a unit membrane. In *Pythium* these apparently bud from a stack of endoplasmic reticulum constituting a Golgi apparatus. However, in most fungi Golgi apparatuses appear to be lacking, and the vesicles then seem to originate from unstacked endoplasmic reticulum. The vesicles would seem to play a key role in hyphal growth. There is convincing evidence that they contain the precursors for wall synthesis. Passing to the apex of the hypha, the vesicles burst not only liberating the precursors in the right place, but also contributing their unit membranes to the necessary extension of the membrane.

Another vital organelle of the protoplasm is the mitochondrion, limited by a double membrane. Mitochondria are involved in the controlled release of energy to power the synthetic processes of the hypha. On passing backwards from the apex, sap-filled vacuoles of increasing size appear, each delimited by a single unit membrane. There may be other particles in the protoplasm such as small lipid bodies.

Fungi grow in nature over or through more-or-less solid substrates or submerged in water. It is impossible to follow the kinetics of their growth there. In the laboratory vegetative growth can be studied in liquid culture and on solid nutrient agar in petri dishes. When either are inoculated with spores, after a variable lag phase during which the synthetic processes become revitalized, they germinate forming a germ tube hypha which grows by the incorporation of material at the tip in a region called the extension zone.

The wall of the extension zone is soft and relatively elastic but increases in rigidity with distance from the tip and becomes completely rigid at the base of the zone. Beyond this point the diameter of the hypha remains constant.

Extension of the germ tube hypha initially occurs at an accelerating rate as both reserves in the spore are mobilized for growth and materials are synthesized in increasing amounts by the absorbing and growing germ tube hypha. After all the reserves are utilized hyphal extension is supported by a constant length of hypha distal to the tip called the peripheral growth zone. Components for wall synthesis, precursors and enzymes, are synthesized in this zone and transported in membrane-bound vesicles to the tip where they are released to form the new wall. At the same time the membrane of the vesicles is incorporated into the membrane of the extension zone. The peripheral growth zone is the maximum length of hypha associated with tip growth. In fungi which form complete septa across their hyphae, it is the length of the apical compartment. Most fungi have septa with pores through the centre and these allow passage of material through. In these fungi the peripheral growth zone is from the tip to the first plugged septum. The length of the zone in part determines the rate of hyphal extension. After a short time hyphae then extend at a constant rate and the rate may be maintained indefinitely if space and nutrients are not limiting. A feature of septate fungi is the presence in the extreme tip of the actively growing hypha of an electron-dense region, the apical body, not delimited by a membrane. As well as being conspicuous in electron-micrographs, it is also visible as a highly refractive particle in living hyphae viewed with the light microscope using phase-contrast. There is strong evidence that this apical body has an important role in controlling the activity of the hyphal apex.

There is a degree of zoning of the organelles in the hyphal tip. Thus in *Pythium* vesicles are especially abundant in the first few micrometres (μm). The first mitochondria are encountered further back and these are ahead of the nuclei. Indeed, in some fungi (e.g. *Flammulina velutipes*) the first nucleus to be encountered may be over 100 μm from the apex.

The extension zone must be capable of utilizing all the precursors produced by the peripheral growth zone so the rate of hyphal extension is also in part dependent upon the width and length of the extension zone. The part of the hyphae behind the peripheral growth zone is still very active in absorbing from its environment and synthesizing new materials, but it does not grow in length. These new materials are used in lateral branch formation. The development of a branch is in reality the creation of a new tip and involves localized softening and partial dissolution of the existing rigid cell-wall. In some fungi (e.g. *Mucor*) branching is initiated only a short distance behind the apex, but in many fungi a lateral develops at a much greater distance from the tip, often below the first cross-wall. For example, in *Flammulina*, in the leading hypha of a colony on agar a side branch is formed about 200 μm from the apex. The local softening of the wall heralding branching is apparently always accompanied by the development of a cluster of vesicles like that in the main apex.

Initially the new branch tip also extends at an accelerating rate as the parent hypha contributes to growth and it starts to absorb, but it soon attains a constant rate as it is supported solely by its own peripheral growth zone. With time more branches are formed and these in turn form branches leading to the formation of a young mycelium in the form of a circular colony.

In young actively growing mycelia in liquid culture, the hyphae are all of the same diameter and total hyphal length is equivalent to the biomass. This is found to increase exponentially as does the number of branches. The ratio between these two properties becomes constant and is called the hyphal growth unit. It is the mean length or more accurately the volume of hypha associated with each branch tip. Growth is exponential because the rate of increase in biomass is a constant function of the biomass present, i.e. all of the mycelial mass is contributing equally to growth. This function is known as the specific growth rate. It can be defined as the mass of fungus in terms of hyphal length or volume produced by a given amount of fungus in a given time. Different environmental conditions will change this specific growth rate. Exponential growth is achieved because nutrients are being assimilated all over the mycelium and are being used to synthesize macromolecules for growth and precursors for wall synthesis while hyphal extension is only occurring at the tips. The number of tips increases by branching to maintain a constant number of tips to hyphal length in the mycelium. The relationship between the rate of hyphal extension and branch formation is an important feature of filamentous growth in fungi. A fungus can regulate the allocation of biomass between hyphal extension and branch formation. In poor nutrient conditions such as when growing through a mineral soil, the resources are concentrated into extension growth and branches are sparse. Thus the fungus spreads rapidly through such barren areas, increasing its chances of reaching new nutrient sources. The hyphal growth unit is increased. Where nutrients are readily available, such as in organic rich pockets in soil, branch formation is profuse such that the available nutrients are utilized. The hyphal growth unit is reduced. Such a strategy gives the filamentous fungi an advantage over unicells.

Theoretically exponential growth can be maintained in continuous liquid culture systems where fresh medium is continually added and old medium is removed at the same rate, but this is seldom true in practice. In theory as a spore grows in liquid culture a spherical colony should develop and initially growth would be exponential but with time, with an increase in volume of the sphere, it becomes more difficult to maintain a surplus of nutrients throughout the colony. Nutrient depletion occurs in the centre and the gradual accumulation of metabolic waste proves toxic and growth slows down. With time autolysis and death occurs beginning in the centre and spreading outwards.

In practice growth is usually measured as an increase in dry mass. If growth is followed in a fixed volume of aerated culture solution of suitable composition, it is found to be slow at first, with an initial lag period, and then the rate increases almost linearly with time. Later growth begins to slow down, partly because of depletion of the limited food supply, but more often because of the accumulation of products of metabolism which check further growth. Specific toxic staling products may accumulate and in addition there may be drastic changes in pH. As the culture ages still further, dry mass begins to fall because of the breakdown or autolysis of some of the components of the hyphae. A typical growth curve is shown in Fig. 2.2.

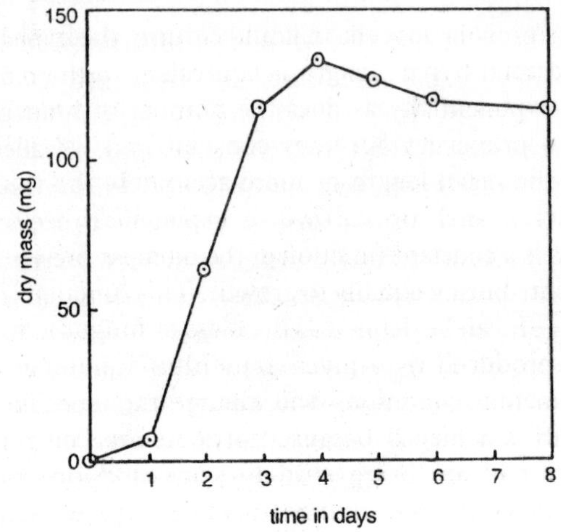

Fig. 2.2. *Fusarium solani*. Growth in aerated liquid medium. (After Cochrane)

Such effects are better observed on solid media on a petri dish. On agar hyphae spread radially from the inoculum developing into a circular colony (Fig. 1.2). After an initial lag phase, a very short exponential phase occurs but cannot be sustained. Nutrient depletion in the centre is more rapid than in liquid cultures. Nutrients can only be replaced by diffusion which becomes rate limiting. Toxic metabolic products again build up, but more quickly, and autolysis of the hyphae can be readily observed. The colony then spreads at a constant linear rate with growth being restricted to the hyphal tips in a narrow annulus of constant width at the colony margin. Again the peripheral growth zone supports this growth. The mean rate at which the hyphae extends is a function of the specific growth rate of the fungus and the hyphal growth unit. Such linear spread is observed in fairy-rings (Fig. 9.3). The rate of radial spread is often used as a measure of growth. Measurements are usually made across two marked diameters and the mean taken. It is a convenient measurement but it is only a measure of extension growth. It takes no account of the density of branching and consequently of specific growth rate.

It should, perhaps, be emphasized that, although apical growth is the rule in the fungal hypha, there are examples of intercalary growth. Notable in this connection is the sub-sporangial growth of the sporangiophore of *Phycomyces* (p. 27) and the elongation of hypha in the growing stalk of toadstools.

In many mycelia anastomosis occurs. Although only rarely seen in fungi without cross-walls, it is an important feature of septate fungi. The process involves the vegetative fusion of a lateral hypha with another nearby. Fusion is apparently always tip to tip. As the apex of a lateral approaches another hypha, it may stimulate it to produce a peg-like outgrowth. The two tips then grow towards each other and fuse, thus establishing open communication (Fig. 2.3). The extent of anastomosis varies with the species and with the conditions of culture. As a result a branched mycelium is converted into a three-dimensional network, thus facilitating translocation throughout the

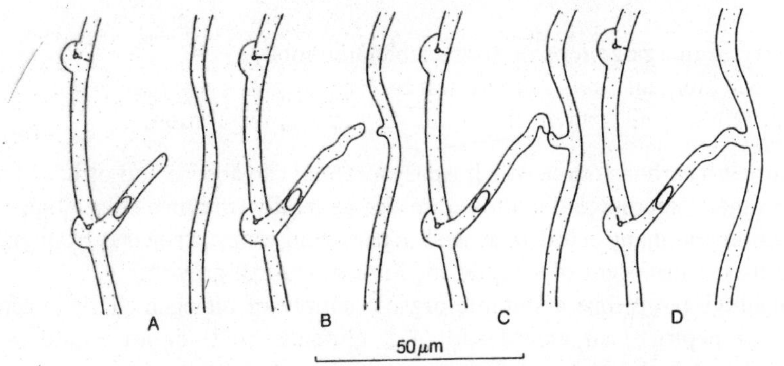

Fig. 2.3. Anastomosis in *Sphaerobolus stellatus*. Four stages (A–D), spread over three hours, in hyphal fusion. In B the lateral from the hypha on the left has induced a short peg-like lateral to develop from the hypha on the right; in C the tips of these two laterals are growing towards each other; and in D they have fused. (After Buller 1933)

mycelium. In a septate fungus such translocation can occur because the cross-walls are not complete. Each septum arises usually as a ring of wall material which on growing inwards stops short of complete closure, leaving a minute hole through which protoplasm can stream.

We may now pass to a consideration of nutrition. The fundamental facts are that, first, unlike green plants, fungi cannot make the needed organic basic food from carbon dioxide and water; and secondly, since their protoplasm is encased in a cell-wall, only substances in solution can be taken in.

In studying nutrition, a fungus is normally grown under sterile conditions in water containing dissolved nutrients. The vessels containing such liquid cultures are usually placed on a shaker so that conditions remain uniform through the volume of fluid in which the mycelium is growing. The medium changes in composition as substances are absorbed and by-products of metabolism are liberated. It would, of course, be preferable to study the growth of a fungus under conditions where the liquid medium is constantly replenished. However, although systems permitting this have been devised, they are complex and mycologists are usually content with continuously shaken cultures of chemically defined, or nearly defined, composition. In nutritional studies a fungus is commonly grown in small, hard-glass, conical flasks, each containing a shallow layer of culture solution, closed with metal caps which, however, allow gaseous exchange. Growth is measured as the dry-mass produced after a given time. However, in some growth studies, especially in those concerned with the effect of external factors, where the precise chemical composition of the medium is not important, agar cultures in petri dishes are used and growth is then measured as an increase in the radius of the colony.

A complete medium for the growth of a fungus must have, in addition to water, certain constituents, namely

(a) a suitable organic substance as a source of carbon
(b) a source of nitrogen

(c) certain inorganic ions in appreciable amounts
(d) other inorganic ions as mere traces
(e) certain organic growth factors in very low concentration.

Glucose is a carbon source which nearly all fungi can absorb and utilize. Two other sugars, sucrose and maltose, are also generally suitable. Most fungi can use starch and many cellulose as well. Apart from water, a sugar is frequently the largest constituent of a liquid medium for fungal growth.

All fungi can utilize a suitable organic supply of nitrogen. This is often given as peptone, an amino-acid (e.g. glutamic acid) or an amide (e.g. asparagine). A limited number can make use of nitrogen in inorganic combination such as an ammonium or nitrate ion, but these are capable of dealing with an organic source as well. Organic substances such as peptone or asparagine can supply both the carbon and nitrogen needs of a fungus.

So far as inorganic nutrition is concerned, potassium, phosphorus (as phosphate), magnesium and sulphur (as sulphate) are all needed in significant amounts in the culture medium. However, calcium, an essential element in the nutrition of green plants, is not apparently needed by all fungi, though it is by some.

There are other inorganic substances of which only traces, to be reckoned in a few parts per million, are required. These include iron, zinc, copper, manganese and molybdenum. They appear to contribute to the structure of essential enzymes. The amounts needed of some of these elements are so minute that a demonstration of their essential nature is difficult. It involves using glassware of high quality, specially cleaned, and salts purified far beyond the limits of most high-grade commercial chemicals. It is not, therefore, surprising that only in a very few fungi is there a clear picture of the 'trace element' requirements.

Much work has been carried out on the needs of fungi for organic growth factors, mainly vitamins, required in minute amounts. For all organisms the same vitamins are necessary. It is a question of whether an external source of a particular one is needed or whether it can be synthesized by the organism. In this respect some fungi are completely self-sufficient and no vitamin need be added to the external medium if growth is to occur. Others require an external source of one or more vitamins. The two most often needed are thiamin (vitamin B_1) and biotin. For example, in the complete absence of thiamin *Phycomyces* (see p. 27) cannot grow. With increase in the concentration of the vitamin in the culture solution from zero to a few micrograms per litre, growth is proportional to the thiamin concentration, but above a certain level increase has no further effect (Fig 2.4). Because of its sensitivity to thiamin, *Phycomyces* can be used to assay the thiamin content of a medium. This biological method is much more sensitive than a purely chemical one. The thiamin molecule is made up of a thiazole half and a pyrimidine half. Where a fungus has a requirement for thiamin, it is sufficient to supply thiazole for the pyrimidine can be synthesized within the hypha.

To support growth, the organic component of a medium need not be in a soluble form, although only substances in solution can be absorbed by a

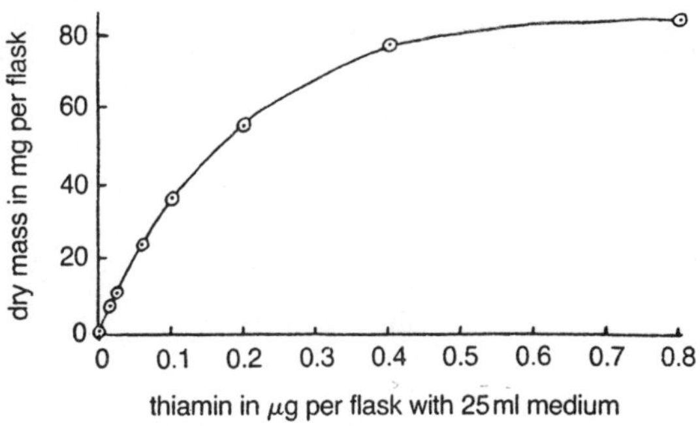

Fig. 2.4. *Phycomyces blakesleeanus.* Dry-mass of mycelium formed in liquid medium containing increasing amounts of thiamin. (Plotted from data by Schopfer in 'Plants and vitamins', *Chron. Bot.* 1943)

hypha. Starch and cellulose are often supplied as the organic food for a fungus in pure culture. Their utilization depends on the production of extracellular enzymes (exoenzymes) which probably pass out from the hyphal tips. Thus starch is converted to soluble glucose under the influence of excreted amylase. Although soluble, sucrose, very frequently used in culture media, does not pass readily into the living hypha. Before absorption, it is normally converted into its component hexoses by extracellular invertase. Important in the physiology of certain fungi, such as *Botrytis cinerea*, which are unspecialized parasites of higher plants, are pectic enzymes which, liberated from the hyphal tips, kill sappy living tissues in advance and break down the middle lamellae of the walls into soluble products. The production of a considerable range of exoenzymes is one of the most characteristic and important features of the fungal mycelium.

Two factors, namely temperature and pH, greatly affect the development of a fungus in culture. In the temperature–growth curve minimum, optimum and maximum values can be recognized. The curve has a characteristic form. It is markedly asymmetric about the optimum, only 10°C usually separating this from the maximum, but around 20°C from the minimum. These cardinal points vary from species to species, although the minimum for most fungi is 2–5°C, the optimum 22–27°C and the maximum 35–40°C. Fig 2.5 illustrates the situation in two toadstools (agarics) that have been frequently used in nutritional studies. In *Coprinus cinereus*, a common fungus on horse dung, the optimum is about 35°C and growth stops below 15°C. *Flammulina velutipes* is abundant on dead elm trunks and is one of the very few toadstools to be found throughout the winter. In this species the optimum is about 25°C and the minimum below 0°C. About 30 thermophilic moulds (e.g. *Thermomyces lanuginosa* and *Mucor pusillus*) are known which fail to grow below 20°C, but

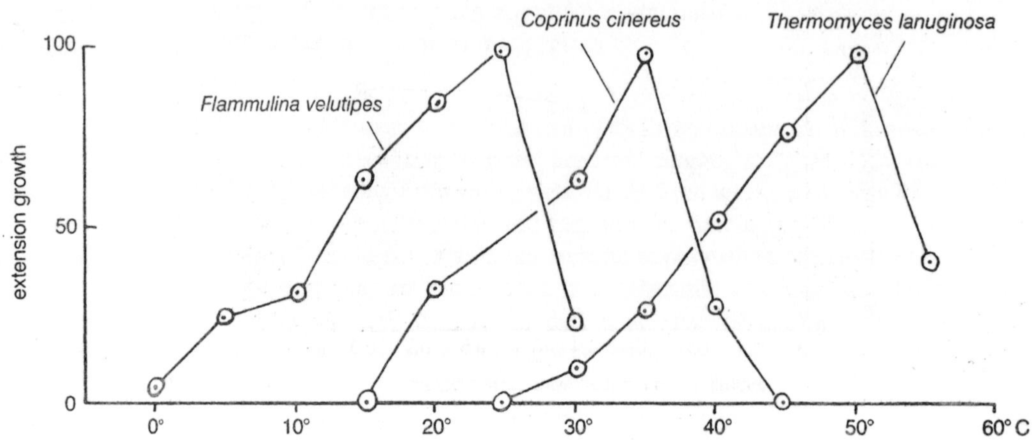

Fig. 2.5. Temperature and growth in three fungi growing on agar. Radial growth for each species is expressed as a percentage of growth at the optimum temperature.

are still capable of growth at 50°C (Fig. 2.5). These have mostly been isolated from such sources as warm compost heaps.

The investigation of the relation between pH and growth presents difficulties. The trouble is that pH cannot be varied without altering other features of the medium. Further, the pH rarely remains constant during growth, but tends to shift, often considerably, even if the medium is reasonably well buffered (Fig. 2.6). Thus, for example, if the nitrogen supply is in the form of an ammonium salt, the cation is absorbed much more rapidly than the anion, and there is a rapid increase in acidity. Again if asparagine is serving as a source of both carbon and nitrogen, it tends to be broken down with the production of ammonia in excess of the nitrogen requirements of the fungus. As a result ammonia remains in the medium which becomes strongly alkaline. In a rigorous experimental study of the effects of pH on growth, it is necessary to adjust the pH of the liquid medium daily by the addition of sterile acid or alkali. Minimum, optimum and maximum values for growth can usually be recognized, but the optimum is often broad, extending over several units of the pH scale. Few fungi grow below pH 3 or above pH 9. The optimum for most is on the acid side of neutrality, often in the range pH 5–6.5. This contrasts with bacteria in which the optimum is usually above pH 7. Mycologists generally use somewhat acid media. Malt agar, so commonly used in routine culture work, is about pH 4.5.

Another factor important for fungal growth is oxygen supply. Nearly all fungi are strictly aerobic, and in the complete absence of oxygen growth ceases. Here again there is a contrast with bacteria, many species of which are anaerobic. However, the actual concentration of oxygen in the air in contact with a fungal culture can be very considerably reduced before any inhibition of growth occurs. *Neurospora crassa*, for example, grows well down to 0.3% oxygen in the atmosphere.

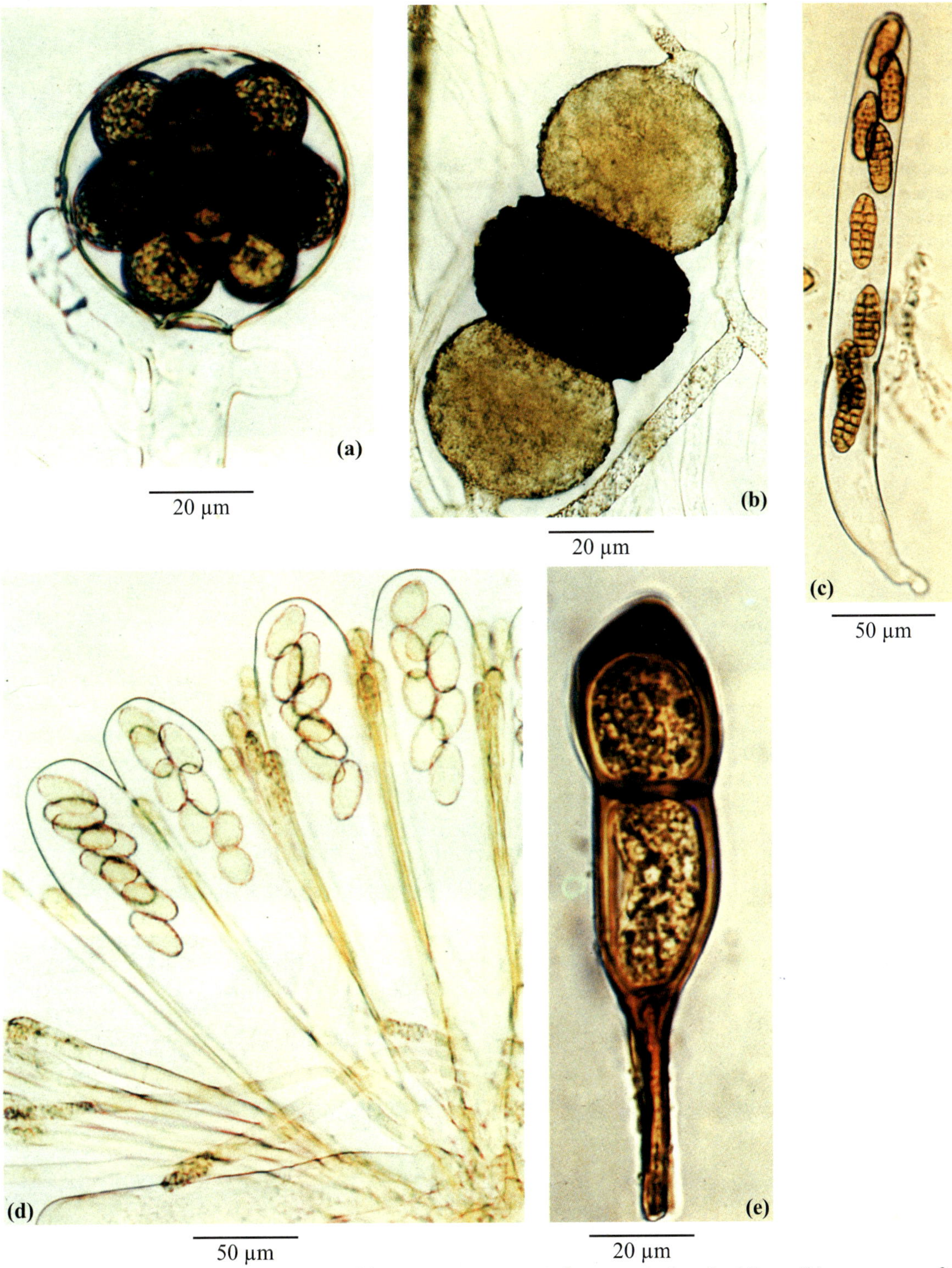

Plate 1 (a) oogonium of *Saprolegnia* with mature oospores and an attached antheridium. (b) zygospore of *Rhizopus* and suspensors. (c) bitunicate ascus of *Pleospora* with inner wall extended and multicellular ascospores. (d) unitunicate asci and paraphyses from a squashed apothecium of *Coprobia.* (e) two-celled teliospore of *Puccinia graminis.*

Plate 2 (a) under surface of a basidiocarp of *Serpula lacrymans* on a dry rotted joist from a house. (b) clump of basidiocarps of *Armillaria mellea*, the honey fungus, arising from a rotted buried root. (c) a cluster of basidiocarps of *Ganoderma adspersum* at the base of a dead beech tree. The brown dust on the trunk and lower basidiocarps are deposits of basidiospores.

(a)

(b)

(c)

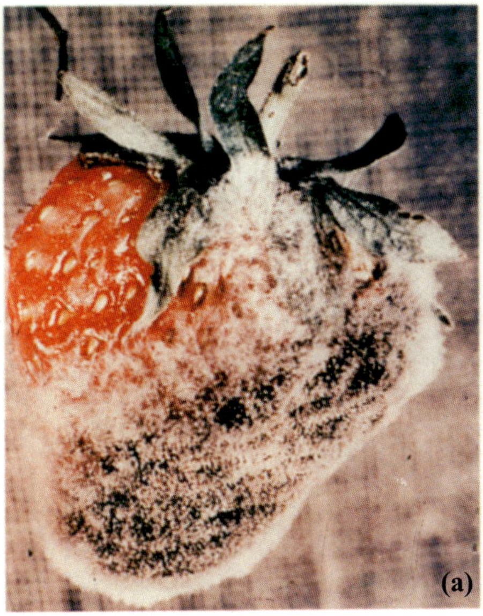

Plate 3 (a) grey mould of strawberries (*Botrytis cinerea*); partially rotted strawberry with conidia already forming. (b) chocolate spot of broad beans (*Botrytis fabae*); discrete necrotic lesions on a leaflet. (c) and (d) brown rot of apples (*Monilinia fructicola*) showing infection via a bird peck; (c) brown rot and initiation of spore production; (d) concentric rings of conidia on a completely rotted fruit. (e) peach leaf curl (*Taphrina deformans*); grossly distorted and highly pigmented leaf.

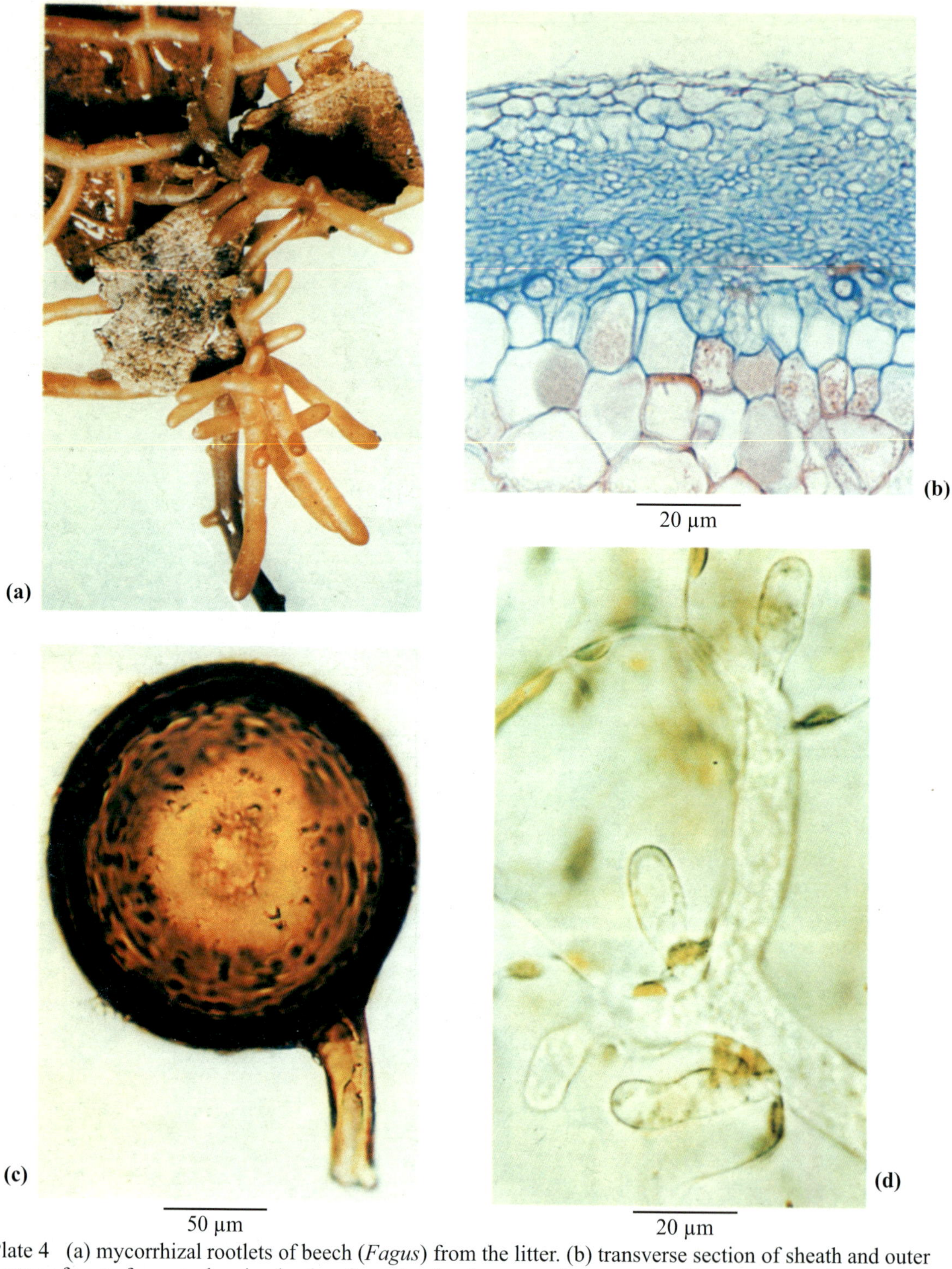

Plate 4 (a) mycorrhizal rootlets of beech (*Fagus*) from the litter. (b) transverse section of sheath and outer cortex of part of a root, showing hyphae from the sheath penetrating as wedge-like projections between the outer cell layer of the root. (c) chlamydospore of *Glomus*. (d) non-septate intercellular hypha and intracellular haustoria of *Bremia lactucae* in a lettuce cotyledon; chloroplasts are still present in the penetrated cells.

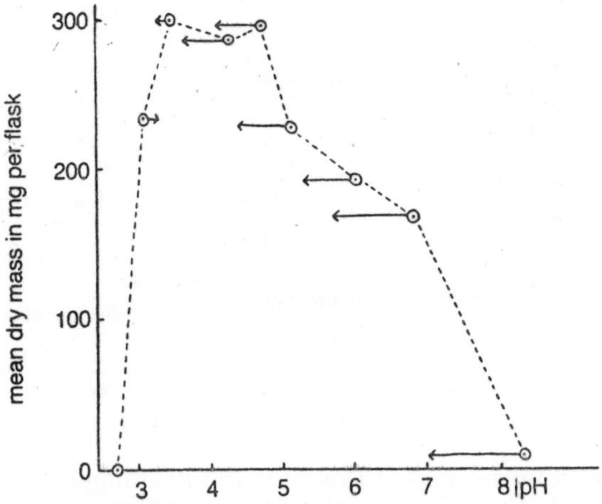

Fig. 2.6. *Phycomyces blakesleeanus.* Dry-mass produced in ten days in flask-cultures with 25 ml liquid medium containing glucose plus asparagine with necessary minerals and adequate thiamin. Each point is the mean of four flasks. Arrows indicate the drift in pH in ten days. (Based on data by Burkholder and McVeigh, *Amer. J. Bot.* **27**, 1940)

As with all living organisms, water is an absolute requirement for fungi and a mycelium can grow only in a watery solution or on an aqueous jelly or in an atmosphere that is nearly saturated. However, many fungi produce spores, reproductive structures or vegetative resting phases capable of surviving, but not of growing, during periods of extreme drought.

During the growth of a mycelium, some substances are being absorbed and others are passing out into the surrounding medium. This inward and outward passage of molecules and ions is known as transport. Reference to this has already been made in connection with the activity of the hyphal tip. Different molecules and ions may pass into the hypha by different mechanisms and by different routes through the membrane. Again the transport of one substance may influence that of another. Mostly transport seems to involve the consumption of energy, although some substances may enter the hypha by simple diffusion. The processes concerned with transport are extremely complex and relatively little is known about them.

Within the mycelium materials may move from one region to another. This is the important process of translocation. For example, when a relatively large basidiocarp is formed above ground, the substances for its construction must have been derived from the mycelium in the nutrient substratum. This particular process can be considered by reference to an experimental study of *Coprinus cinereus*. The increase in dry mass was followed having regard separately to the vegetative mycelium and to the aerial basidiocarps (Fig. 2.7). It is clear that for the production of the latter, there must have been

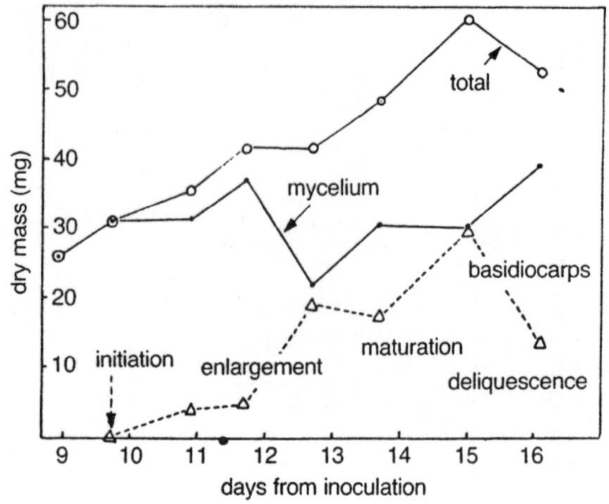

Fig. 2.7. *Coprinus cinereus*. Dry-mass of total fungal material of vegetative mycelium and of basidiocarps plotted against time. Based on samples taken at intervals from a population of pure cultures. Each point for each sampling time is the mean of four replicates. In the lower part of the figure the stages in basidiocarp development are indicated. When the basidiocarps deliquesce, there is a slight fall in total weight. (Constructed from data in Madelin, *Ann. Bot.* **620**, 1956)

considerable translocation of material from the mycelium in the nutrient substratum.

A split-dish method has been used by several workers to provide information about translocation in the mycelium. This technique may be illustrated by a specific example (Fig. 2.8). In this experiment agar in the petri dish was split into a central region containing radioactive phosphate (^{32}P) separated by two narrow agar-free ditches from agar on either side lacking labelled phosphate. The dish was inoculated in the centre with *Rhizopus* (see p. 28). The growing fungus had no difficulty in bridging the ditches which, however, acted as diffusion barriers to the radioactive phosphate in the middle region. When the mycelium had completely filled the dish, sample areas (1–14) were removed and the amount of radioactivity in each was determined. It is clear (Fig. 2.8C) that the ^{32}P was translocated to the periphery of the dish, the considerable build-up of the tracer there probably resulting from the higher water loss (transpiration) from the fungus in that region.

Translocation in the stalk of toadstools has also been studied. Thus in cultures of *Flammulina velutipes*, it was found that ^{32}P, injected into the agar near the base of a basidiocarp, appeared in the cap 1½ hours later.

The construction of the fungal mycelium seems well suited to translocation. In many of the simpler fungi such as *Mucor* or *Phycomyces*, the absence of cross-walls allows movement to occur from one part of the mycelium to another. Further, in septate fungi, not only has each septum a central pore through which streaming can occur, but also anastomoses provide short-cuts for translocation across the mycelium. It is to be noted, however, that unlike

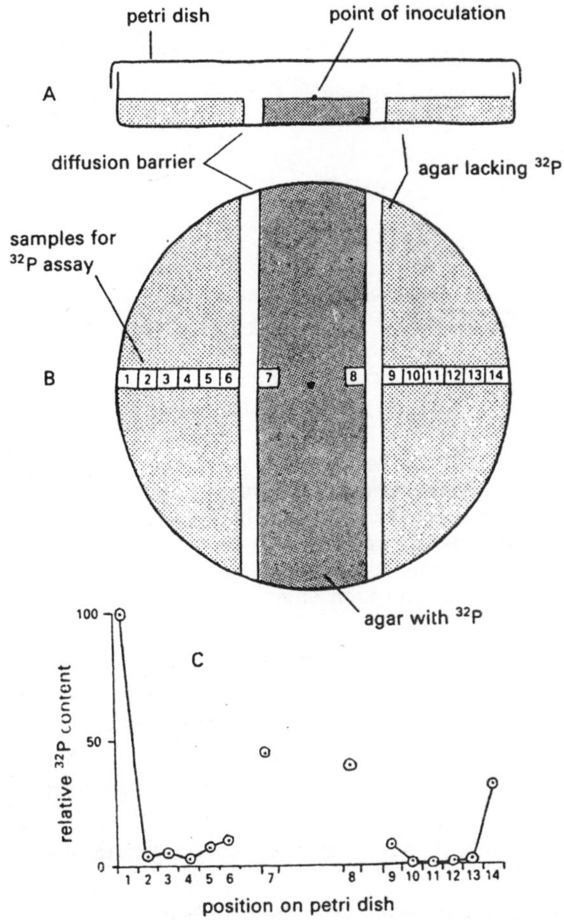

Fig. 2.8. Split-dish method of studying translocation. A, vertical section of petri dish with agar in middle having ^{32}P; outside without ^{32}P separated by diffusion barrier; dish inoculated in the centre has given a colony of *Rhizopus* extending over the whole dish. B, surface view of the same indicating the pieces of agar, including the fungus, which were analysed for ^{32}P. C, result of analysis showing translocation to the periphery of the dish. (Based on Lucas, *Trans. Brit. Mycol. Soc.* **69**, 1977)

the position in higher plants, there are no separate channels concerned with translocation of water and of organic substances.

Much remains to be discovered about the mechanisms of translocation in fungi. In mucoraceous species transpiration clearly plays a part. This is easily demonstrated in a petri-dish culture of *Phycomyces* (see p. 27) viewed through the base under the low power of the microscope. As the mycelium ages it becomes simplified by the occlusion of the smaller branches. In each main hypha there is a thin, static layer of protoplasm around a central zone occupied by protoplasmic masses separated by clear, vacuolar regions. If an

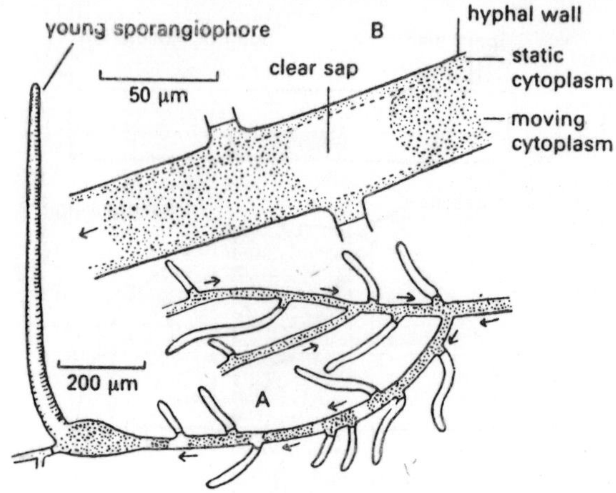

Fig. 2.9. *Phycomyces blakesleeanus*. A, precise drawing of part of a mycelium on agar. Many empty laterals are cut off with cross-walls. The observed flow in the mycelium is shown by arrows. Flow leads to the base of a sporangiophore. The sporangiophore here figured is young. Being surrounded by air its contents cannot be seen and it is shown in solid view. B, small part of the mycelium much enlarged.

open dish is used, so that transpiration can occur, movement of material in the central region of the hypha is easily seen, and if a particular stream is followed, it always leads into the base of an aerial sporangiophore (Fig. 2.9). Rates of flow of 20–40 μm/s are usual. The flow seems to be powered by evaporation from the end of the sporangiophore, the rest being protected from water loss by an outer, lipid layer. If a culture in a covered petri dish is observed, streaming is seen almost to have stopped as transpiration is reduced to near zero.

The movement of material in a septate mycelium is not obviously related to transpiration. Streaming from cell to cell through the pores in the septa seems always to be towards the growing apex. It has been suggested that the development of vacuoles in the older regions of a hypha is responsible for driving protoplasm towards the hyphal tip (Fig. 2.10).

So far nutrition has been considered without reference to the nature of growth or whether vegetative mycelium or reproductive structures are being produced. The nature of the medium and other factors of the environment may greatly affect the type of growth that occurs. Many researchers have studied the physiology of sporulation (reproduction) in fungi. It is difficult to generalize on this matter, but in the main conditions for reproduction tend to be more exacting than are those for vegetative growth.

So far as temperature is concerned the cardinal points for reproduction may be different for mycelium growth and for sporulation. For example in *Sphaerobolus* (see p. 98) the optimum for basidiocarp production is around 15°C, but for vegetative growth it is 30°C at which temperature no basidiocarps are formed (Fig. 2.11).

Mycelial growth and reproduction may have different vitamin require-

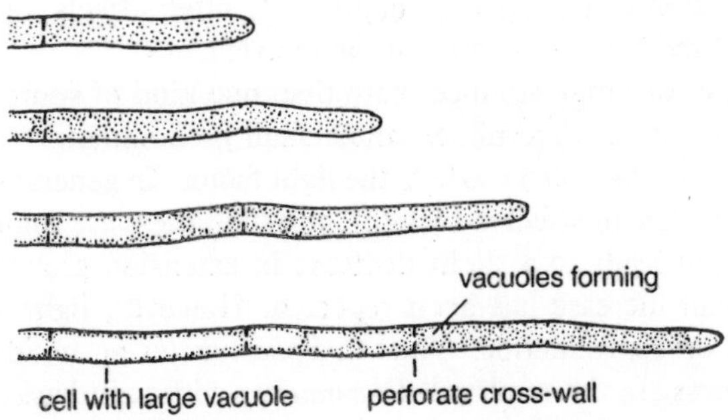

Fig. 2.10. Diagram of four successive stages in the growth of a hypha showing how vacuole development and enlargement may drive cytoplasm towards the apex. Part of the cytoplasm in the apical cell is derived in this way, and part is newly synthesized.

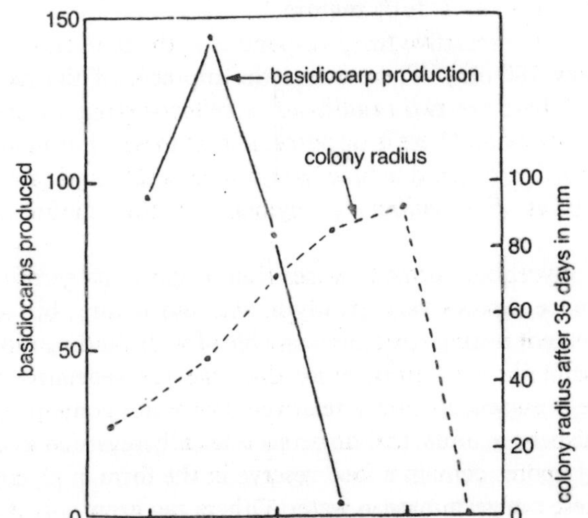

Fig. 2.11. *Sphaerobolus stellatus*. Growth and basidiocarp production in relation to temperature. Mycelial growth expressed as radius of the colony after 35 days. Basidiocarp production: cultures initially in darkness for two weeks at 20°C, then transferred to continuous white light (1000 lux) at different temperatures. (After Alasoadura, *Ann. Bot.* **27**, 1963)

ments. Thus in *Sordaria* (p. 61), a fungus that has been widely used in nutritional studies, an external supply of thiamin is needed for vegetative growth, but if perithecia are to be formed, biotin must also be present in the medium.

The concentration of the organic food supply often affects reproduction, which is usually reduced if the medium is too rich.

A large number of fungi produce more than one kind of spore, and often nutrition or temperature determines which shall predominate.

So far no reference has been made to the light factor. In general the amount of vegetative growth is unaffected by light, although there are some species in which illumination leads to a slight decrease in extension growth, and still fewer in which an increase has been reported. However, light often has a profound effect on reproduction. Thus in *Sphaerobolus* no basidiocarps are formed in darkness. In the toadstool *Flammulina*, although basidiocarps are initiated in the dark and stalks are formed, cap expansion depends on light. Many moulds produce their conidia only in light. In a culture of such a fungus grown under natural lighting conditions, there is a striking zonation with circles of spore-bearing mycelium formed during the day, alternating with circles free from spores produced at night. This is particularly obvious with a mould having coloured spores (e.g. *Penicillium*).

In addition light may have a directional effect. The sporangiophores of *Phycomyces* and *Pilobolus* are positively phototropic, as are the stalks of many toadstools before the cap is fully mature.

The light to which sensitive fungi respond is at the blue end of the visible spectrum (below 480 nm). There arises the problem of the nature of the photoreceptor. There are two candidates: a yellow–orange carotenoid, or a flavin such as riboflavin. There is no agreement as to which is involved. Most fungal structures that respond to blue light possess a coloured carotenoid, but in some, such as *Flammulina*, a pigment of this nature cannot be demonstrated.

Since a new mycelium normally arises from a spore, its germination is of special importance. Spores vary greatly in size and nature, but all agree in being to some extent resting structures capable of withstanding more extreme conditions of drought and temperature than can the vegetative mycelium. This resistance is associated with a relatively low water content. In addition most are also dispersive units, and dispersal is usually regarded as their major function. Many spores contain a food reserve in the form of glycogen or fat, and some of these can germinate in water. Others can grow only if there is an external supply of food. For many it is difficult to induce germination. Often heat-shock is required. *Rhizina inflata* is a discomycete fungus attacking the roots of conifers. Trees are particularly affected near old sites of brushwood fires, and there the fungus produces its ascocarps. It seems that ascospores require exposure to around 40°C before germination can occur. However, different methods may be required to break dormancy in other fungi. Thus in the urediniospores of some rusts, immersion in water is necessary to allow the leaching of an inhibitor. In many of the larger fungi, especially amongst Gasteromycetes and in species of *Lactarius* and *Russula*, mycologists have not succeeded in germinating the spores, presumably because a set of conditions is needed which has not yet been discovered.

Some mycologists have made detailed studies of the biochemical changes that occur during germination. There has been considerable work with spores

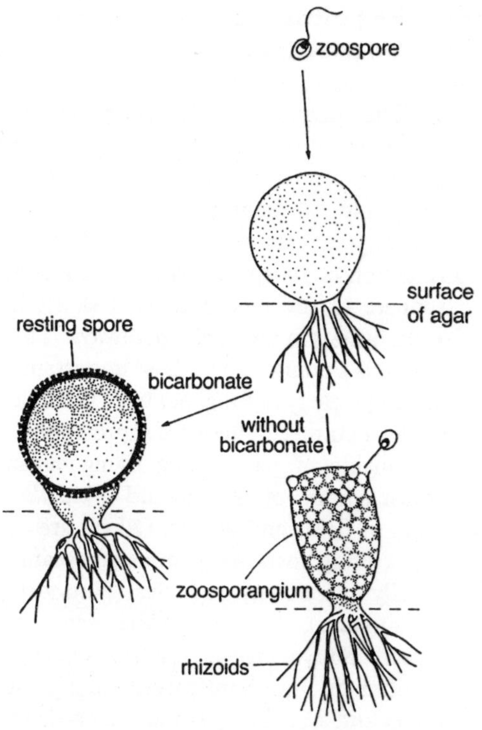

Fig. 2.12. *Blastocladiella emersonii.* Diagram of development showing shunt to resting-spore condition produced by bicarbonate.

of *Rhizopus* and *Phycomyces*. It seems that once a spore is committed to germination, the machinery of metabolism is mobilized in an orderly manner starting with the synthesis of protein and RNA.

In this brief review of the physiology of fungi, emphasis has been on the effect of the environment on overall growth. However, once the fungus has absorbed the necessary raw materials from the surrounding medium, there remains the enormous and complex problem of metabolism. There are elaborate interrelated biochemical processes in the living hypha each catalysed by a particular enzyme produced under instructions from the DNA of the nucleus. On the one hand there are syntheses leading to the production of nucleic acids, proteins, fats, glycogen and cell-wall polymers, and, on the other hand, the breakdown of compounds, particularly carbohydrates, with the stepwise liberation of energy some of which is channelled back into synthetic reactions. In this book, we are, however, largely concerned with structure particularly in relation to function, and with the fungi as related to their environment. It is worth remembering, however, that the form of a fungus is the result of its metabolism, and it is natural, therefore, that some mycologists should have attempted to relate morphology with biochemistry.

Such an attempt has been made using the microscopic aquatic species *Blastocladiella emersonii* (Fig. 2.12). When the zoospore settles, fine branched 'rhizoids' are formed which penetrate the nutrient medium, the body of the spore enlarging either into a colourless, thin-walled sporangium, from which in due course numerous zoospores are set free, or into a resting-spore with a thick, pigmented wall. The fungus is of limited growth so that a culture consists of enormous numbers of individuals each derived from a zoospore. Mass synchronized cultures can be produced, so that for any stage of development morphology and biochemical activity can be simultaneously studied. Provided a critical stage has not been passed, development can be switched from the production of zoosporangia to resting-spore formation by the addition of bicarbonate to the medium. This shunt has been clearly associated with precise alterations in metabolic pathways. This classic analysis is in no way complete, but it is a pointer towards an important area of research.

Inevitably work on the physiology of fungi has been possible only with those that grow readily in artificial culture. These are saprotrophs. Some fungi are obligate saprotrophs incapable of parasitizing living organisms. Familiar examples are *Mucor*, most species of blue mould (*Penicillium*) and a great number of the large fungi of fields and woods. Others are obligate parasites capable of growth only on a living host. Many important plant parasites belong to this category, especially the downy mildews (Peronosporales) and the powdery mildews (Erysiphales). Between these two extremes are all gradations. The grey mould (*Botrytis cinerea*) is a frequent saprotroph on dead organic matter, but it is often an important parasite of a great range of sappy tissue such as lettuce leaves and strawberry fruits. Towards the other extreme is *Phytophthora infestans*, the cause of potato blight, which in nature occurs only as a parasite, but can be grown in the laboratory as a saprotroph on oatmeal agar.

Zygomycotina and Mastigomycotina 3

In this and the following three chapters the structure and function of fungi will be considered in a taxonomic framework. Systems of classification differ in their detail from authority to authority. Many mycologists have divided the fungal kingdom into five divisions: Zygomycotina, Mastigomycotina, Ascomycotina, Basidiomycotina and Deuteromycotina. In this book we shall consider representatives from the more important classes within each of these. The Ascomycotina, including the lichens, is by far the largest division in terms of numbers of species (Fig. 3.1).

It is natural to speculate on possible phylogenetic relationships between the classes of the fungi. For reasons which will be discussed later, Deuteromycotina can be left out of this consideration. So far as the other classes are concerned, their phylogeny is obscure, and this obscurity is enhanced by the extreme paucity of the fungal fossil record. The plain truth is that there are no reliable indications of affinity. Members of the Zygomycotina and Mastigomycotina appear to be relatively simple compared with other fungi and are often referred to as 'lower fungi', a term conveying a suggestion that the other divisions, which constitute the 'higher fungi', have evolved from them. Some mycologists like to think that the Basidiomycotina have arisen from the Ascomycotina but this thesis, although attractive, can bring little evidence to its support. Phylogenetic speculation is, perhaps, best avoided, except within very narrow limits.

ZYGOMYCOTINA

Zygomycetes constitute a class within the Zygomycotina of rather inconspicuous fungi reproducing asexually by non-motile spores each surrounded by a spore wall. The sexual process is rather characteristic involving the fusion of two gametangia to form a thick-walled zygospore. Of the three orders recognized in the class, the largest is Mucorales.

Mucorales

This is a widespread group of moulds occurring as furry growths on damp bread, rotting fruit, and especially on dung in humid weather. Although not

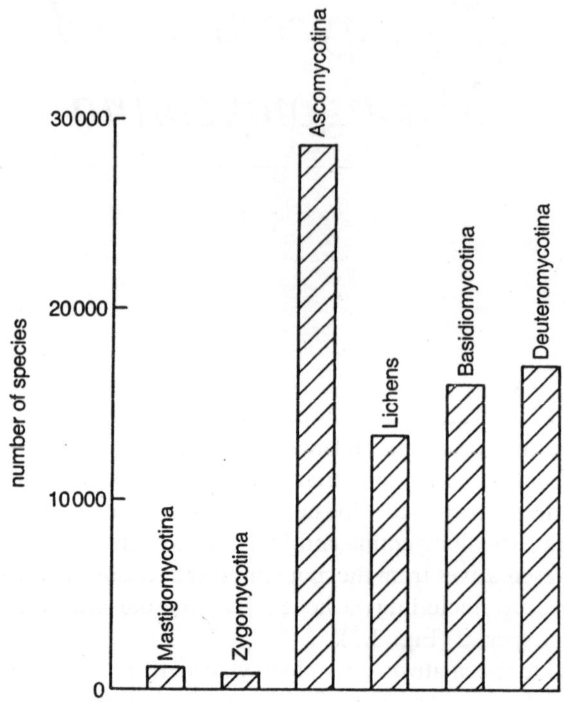

Fig. 3.1. The five major divisions of fungi showing the approximate numbers in each (the lichens and Ascomycotina are shown separately).

directly visible, these fungi are important saprotrophs in soil. Some species have the peculiar habit of parasitizing other members of the order.

The mucoraceous moulds grow readily and rapidly in pure culture. Many species are ideal organisms for laboratory study and they have been extensively used in mycological research, particularly the genera *Mucor*, *Phycomyces* and *Pilobolus*. In the first chapter reference has already been made to the asexual cycle of *Mucor*. The branched mycelium is usually without cross-walls when young, but scattered septa often develop as it ages. Frequently a lateral hypha is occluded by a cross-wall. Further walls are formed in connection with reproductive structures.

In *Mucor* and *Phycomyces* long, and usually unbranched, hyphae (sporangiophores) grow into the air. As development proceeds the tip of the sporangiophore swells into the incipient sporangium. In this cytoplasm and nuclei accumulate, and then the sporangium is delimited by a septum, strongly arched from its inception, which bulges into the sporangium as the columella. Within the sporangium the multinucleate mass of protoplasm cleaves into a large number of equal portions, in some species uninucleate but in most with several nuclei. These portions are the young spores (sporangiospores) each of which develops a surrounding wall (Fig. 3.2).

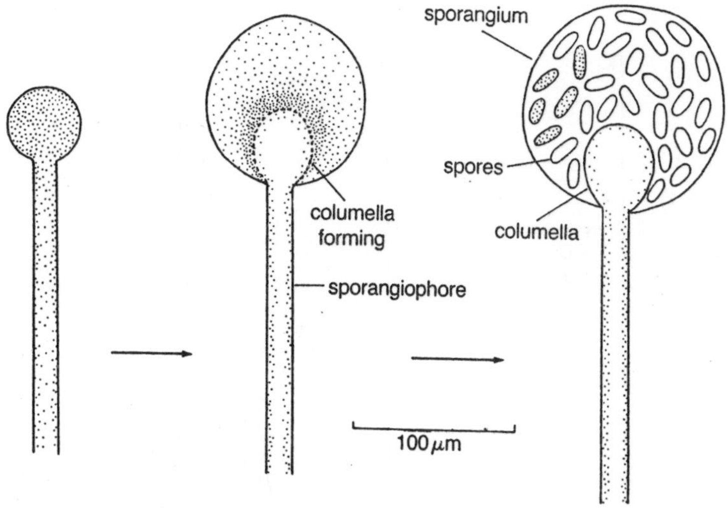

Fig. 3.2. *Phycomyces blakesleeanus.* Stages in the formation of a small sporangium.

A large sporangium of *Phycomyces* may contain 100 000 spores, but in culture tiny sporangia are also formed. The thin wall of the sporangium in species of *Mucor* and some other genera is ornamented by needle-like crystals of calcium oxalate, but in *Phycomyces* the wall is smooth.

A culture of *Mucor* or *Phycomyces* with its long erect sporangiophores sometimes looks like a forest of pins, and, indeed, mucoraceous fungi have been called 'pin moulds'.

In *Phycomyces* the sporangiophores are especially long, often around 100 mm. They have been used extensively in the study of phototropism, hyphal growth and cell-wall structure. The aerial sporangiophore at first has normal apical growth. Then, following a static period while the sporangium is forming, growth continues by the stretching of a 2 mm zone immediately below it. Not only does the sporangiophore elongate, it also rotates. Rotation seems to be related to the fact that the microfibrils in the wall are laid down in a condensed helix. It has been suggested that the stretching involved in the elongation of the sporangiophore inevitably leads to rotation, in much the same way that rotation occurs when a vertical coiled spring secured at the top is extended by a weight attached to its free, bottom, end (Fig. 3.3).

Once the sporangium is mature, there arises the problem of the dispersal of its spores. In most species of *Mucor*, the sporangial wall, except for a collarette around the base of the columella, dissolves. It is said to be 'diffluent'. Water apparently passes into the spore mass through the columella to form a sporangial drop (Fig. 3.4C). When exposed to air the drop dries and the spores become firmly cemented to the columella and cannot be blown off even by quite strong winds. In *Phycomyces* the sporangium wall is extremely thin, but it is not diffluent. On contact with a hard object, the wall ruptures and the

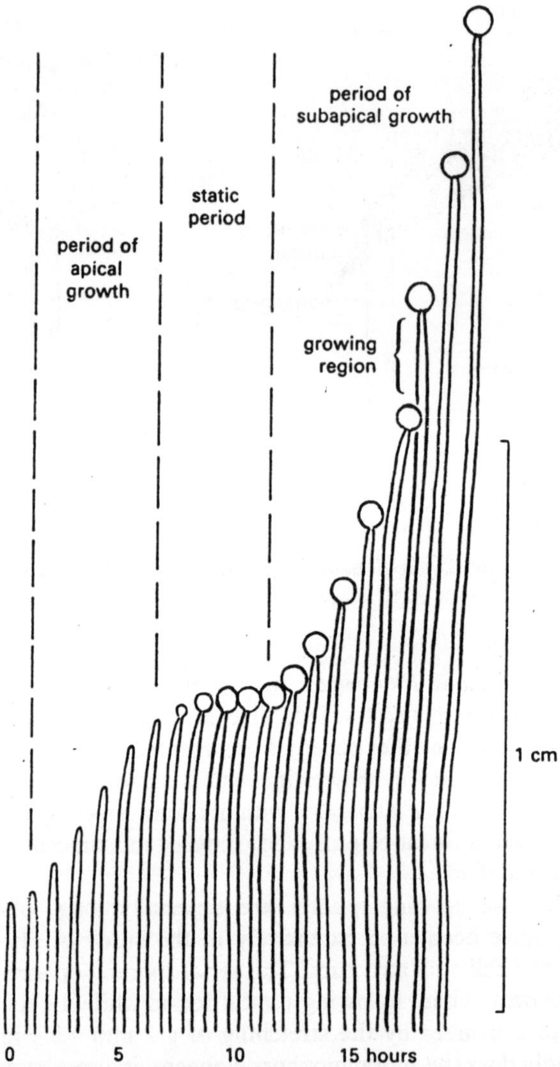

Fig. 3.3. *Phycomyces blakesleeanus*. Same sporangiophore at hourly intervals. (Traced from photographs by Delbruck *Ber. d. d. Gesell.* **25**, 1963)

slimy mass of spores oozes out. In *Phycomyces* and most species of *Mucor*, direct wind dispersal of spores does not seem possible. However, in a few species of *Mucor*, such as *M. plumbeus* (Fig. 3.4D) no sporangial drop is formed, neither is the wall diffluent. At maturity the sporangium is converted into a dry mass of spores and wall fragments easily dispersed by air currents.

One of the commonest mucoraceous moulds is *Rhizopus stolonifer*, abundant on damp bread and rotting fruit. The sporangia appear black. This mould produces fast-growing aerial hyphae (stolons) which arch over and touch down on a solid surface ahead of the main mycelium (Fig. 3.5) Close to its point of contact the stolon forms a tuft of root-like hyphae (rhizoids). It

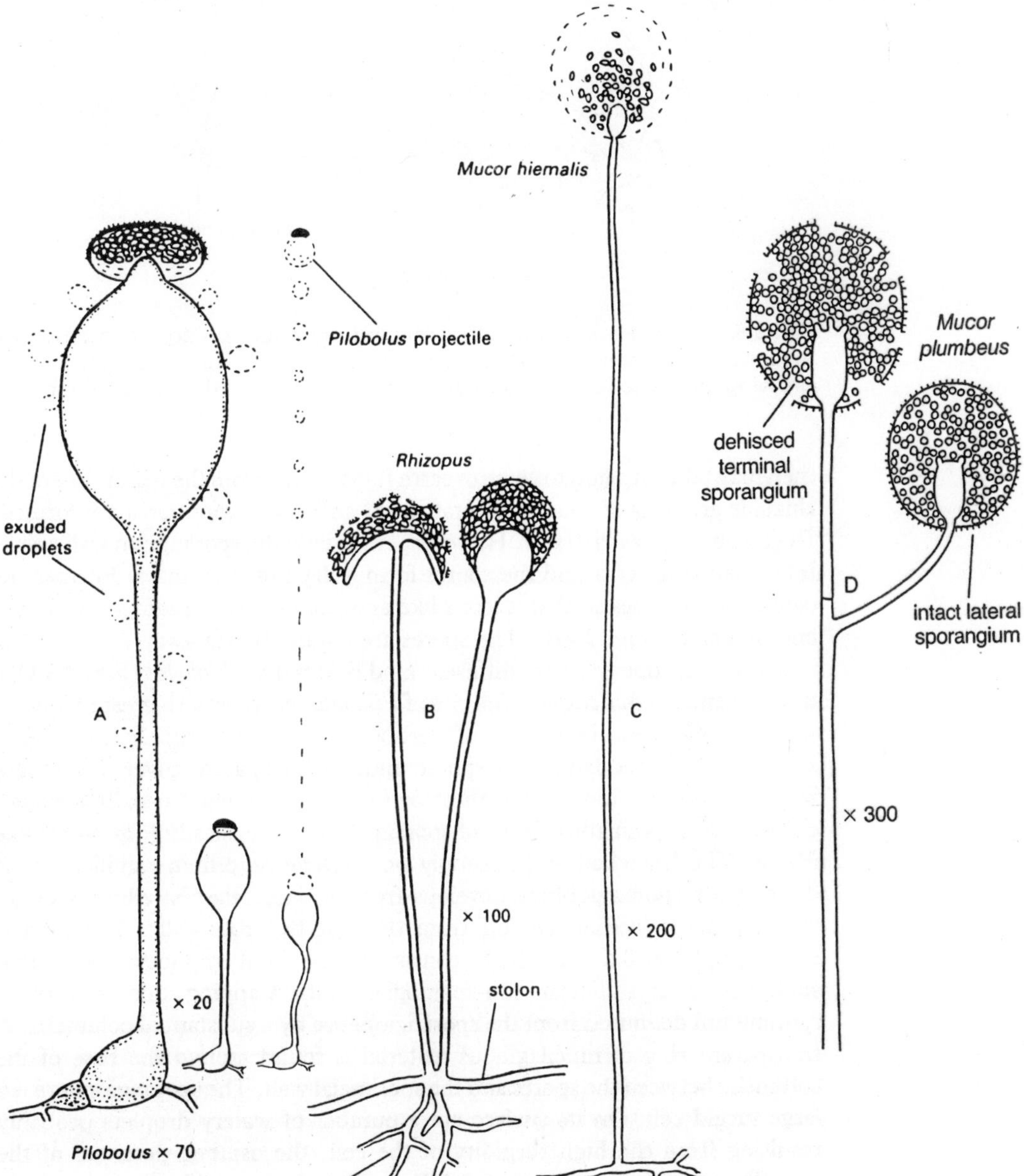

Fig. 3.4. Sporangiophores of Mucoraceae shown in longitudinal section. A, *Pilobolus kleinii*: sporangiophore (× 70) and small-scale diagram of discharge. B, *Rhizopus stolonifer*, pair of sporangiophores; one with a ripe sporangium, the other after drying (× 100). C, *Mucor hiemalis*, the sporangium has become converted into a sporangial drop (× 200). D, *Mucor plumbeus*, one sporangium mature and undehisced; the other has dehisced (× 300).

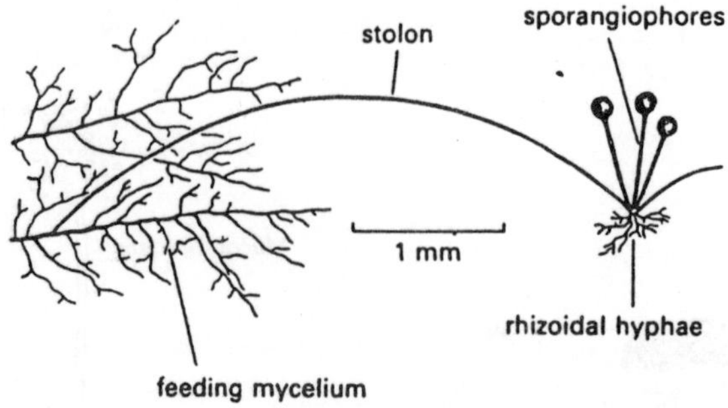

Fig. 3.5. *Rhizopus stolonifer.* The feeding mycelium on agar has produced an aerial stolon which has touched down again on the agar in advance of the growing colony to form a group of rhizoids and three aerial sporangiophores.

then may take off again to come to earth further on. From the region where the rhizoids are formed, one to several short, stiff sporangiophores are formed. Development is as in *Mucor*. However, eventually the sporangium wall cracks into small fragments and the spores form a dry powdery mass. Further the columella collapses so that it looks like an upside-down pudding-bowl at the end of a stick (Fig. 3.4B). The spores are readily blown away.

Spore liberation of a very different kind is found in *Pilobolus* (Fig. 3.4A), a unique genus in this respect. Species of *Pilobolus* develop with great regularity when freshly deposited dung of herbivorous animals (e.g. horse, rabbit, sheep) is kept in a damp atmosphere under a transparent cover in the light with adequate aeration. After three or four days the dung usually becomes covered with a miniature forest of sporangiophores. Basically *Pilobolus* is like *Mucor*. The branched and generally non-septate mycelium ramifies in the dung. Each sporangiophore develops from an intercalary swelling (trophocyst) of the mycelium cut off from the rest by cross-walls. The mature sporangiophore, 0.5–1 cm high, consists of the basal trophocyst, a straight stalk and a crystal-clear sub-sporangial bulb. Capping this is a black sporangium delimited from the sporangiophore by a substantial columella. A transparent ring of mucilaginous material is found around the base of the columella between the spores and the sporangial wall. The sporangiophore is a large turgid cell. On its surface are a number of watery droplets probably resulting from the high turgidity of the cell, the osmotic potential of the vacuolar sap being about 0.7 MPa. The sporangiophore is strongly phototropic and the longitudinal axis of the subsporangial bulb is accurately aligned to the incident light. Lining the wall of the sporangiophore is a thin layer of protoplasm surrounding a large vacuole. However, where the stalk joins the subsporangial bulb the protoplasm is thicker almost, but not quite, bridging the vacuole. The protoplasm here is rich in minute oil drops, orange–yellow with dissolved carotene. The orange region below the crystal bulb is clearly visible with the unaided eye. It has been argued convincingly that the upper part of the sporangiophore of *Pilobolus* is a simple eye (ocellus) having a

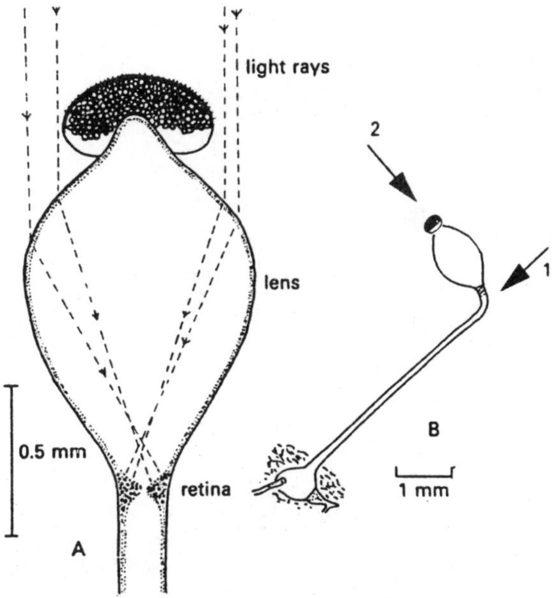

Fig. 3.6. *Pilobolus kleinii.* (A) Upper part of sporangiophore acting as a simple eye; (B) sporangiophore originally developed in light from direction 1, but two hours ago illumination was altered to direction 2.

clear lens and an orange retina. If the retina is evenly illuminated by light focused by the lens, there is no reaction. If, however, it is not symmetrically lighted, curvature of the sporangiophore rapidly occurs just below the bulb until even lighting results (Fig. 3.6). There is a fresh crop of sporangiophores daily and discharge of sporangia occurs around noon. At the moment of discharge a sporangiophore ruptures around a circular line of weakness just below the columella and the whole sporangium is squirted to a distance of a metre or more towards the incident light. On striking a grass leaf or other surface, it becomes orientated in such a way that the black sporangial wall faces outwards, protecting the spore from desiccation and ultra-violet light. The mucilage, as it dries, cements it down onto the herbage where it remains until eaten by a browsing animal. The spores then pass through the animal not only unharmed, but also experiencing conditions in the alimentary canal that stimulate subsequent germination in the dung. *Pilobolus* is only one of a large and beautiful group of coprophilous species of diverse taxonomic nature. Nearly all are dispersed by passage through the animal. Further, it is by no means the only mucoraceous member of this dung fungal flora, but it is unique amongst these in the violent discharge of its sporangium.

Thamnidium (Fig. 3.7) is a genus of the Mucorales occurring commonly on the droppings of field mice but it has also been recorded from soils and even meat in cold storage. In addition to a *Mucor*-type terminal sporangium, the long sporangiophore bears numerous minute sporangia (sporangioles) on short, highly-branched laterals. The large sporangium behaves like that of

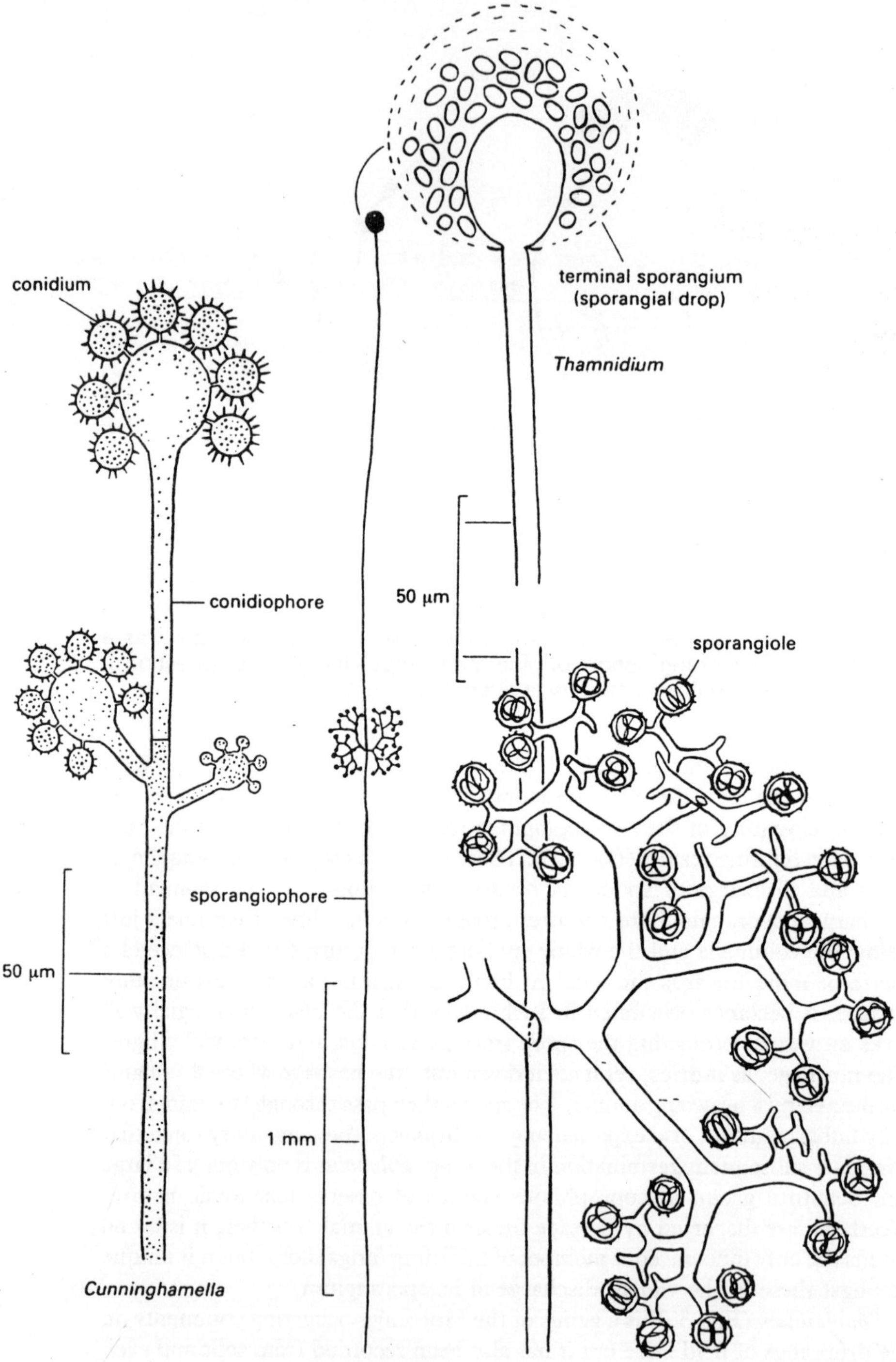

Fig. 3.7. *Cunninghamella*: conidiophore with conidia; and *Thamnidium*: sporangiophore with a sporangium and sporangioles.

Mucor, but the sporangioles, each with only three or four spores, are readily detached and blown away by wind.

In species of *Cunninghamella* (Fig. 3.7), a common saprotroph in soil, the spores are conidia. 'Conidium' is a general term for a spore produced externally. There are different kinds of conidia almost certainly with different evolutionary histories. The conidium of *Cunninghamella* is usually interpreted as a one-spored sporangiole in which the sporangium wall and the spore-wall are firmly united. The sporangiole of *Thamnidium* is, indeed, a big step towards a conidium.

It is interesting to note in passing that several separate evolutionary lines in Mucorales have been recognized each leading from sporangium to conidium. A conidium is more easily freed for aerial dispersal than is a spore developed within a sporangium. There would thus seem to be evolutionary pressure towards the development of conidia.

Some mucoraceous moulds produce chlamydospores, but in others (e.g. *Phycomyces*) they are absent. They are formed in an intercalary position as short, somewhat swollen portions delimited on either side by cross-walls. The chlamydospore is a resting-spore with a firm, resistant wall and a store of fatty reserve. In *Mucor racemosus* these spores are particularly abundant, occurring not only in the vegetative mycelium, but also in the sporangiophores (Fig. 3.8).

A striking feature of certain species of *Mucor* is the capacity to assume a yeast-like state when grown in a liquid medium with a high sugar content. Thus *M. racemosus* under these conditions not only buds like a yeast (Fig. 3.9), but also ferments the sugar with the production of alcohol.

It is now necessary to consider the sexual stage in Mucorales which is highly characteristic. Two more or less equal gametangia fuse to form a black, thick-walled, warty zygospore. In most species of *Mucor*, cultures derived from single sporangiospores fail to form zygospores. However, it has been found that if, in a particular species of *Mucor*, a large number of single-spore cultures are obtained from different sources they could be sorted into two types or strains, called 'plus' and 'minus'. The 'plus' and 'minus' cultures were morphologically alike. If in a culture dish a 'plus' colony met another 'plus' one, or if a 'minus' encountered another 'minus', there was no sexual response. However, if 'plus' and 'minus' met, zygospores were formed in the zone of contact. A species with 'plus' and 'minus' strains is said to be heterothallic, and the phenomenon is known as heterothallism (Chapter 8).

In a heterothallic *Mucor*, if 'plus' and 'minus' spores are planted a few centimetres apart on the same agar surface in a petri dish, each produces a circular colony. Both colonies form sporangia, and all the spores from all the sporangia of one colony are of the same sign. When the margins to the two colonies approach, but before they actually meet, each stimulates the other to produce long, aerial hyphae called zygophores which arch over towards the opposite colony (Fig. 3.10). Zygophores from the two colonies of different mating type come in contact, often laterally, progametangia are produced, and zygospore formation follows (Fig. 3.10).

As the progametangia enlarge, each forms a wall which delimits a multinucleate gametangium, the residual part being the suspensor. Then the

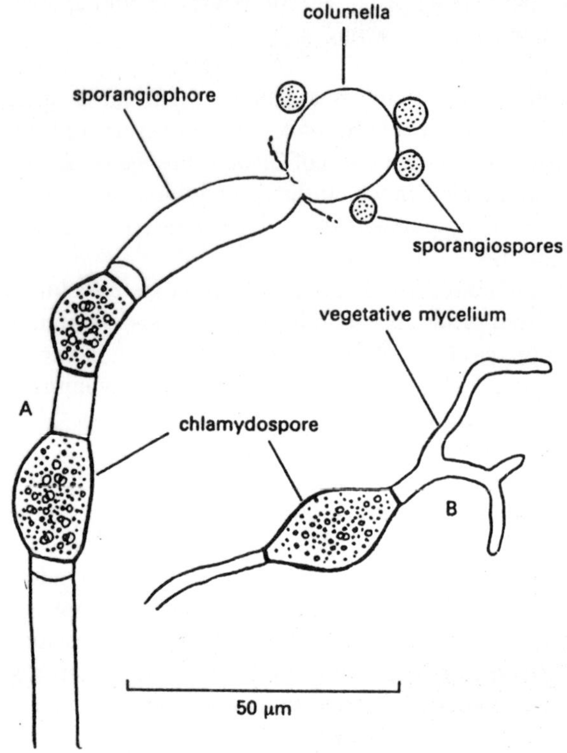

Fig. 3.8. *Mucor racemosus*. A, sporangiophore with dehisced sporangium and four sporangiospores still adhering to the columella; two chlamydospores have formed in the sporangiophore. B, side branch of the vegetative mycelium with an intercalary chlamydospore.

wall separating the two gametangia dissolves from the centre outwards, and their contents intermingle. The resulting fusion cell is the young zygospore. The available cytological evidence, which is not very convincing, suggests that the nuclei associate in pairs and fuse. It is highly probable that in each fusing pair one is 'plus' and the other 'minus'. The young zygospore now enlarges and accumulates a considerable amount of fatty food reserve. The original walls of the gametangia remain thin and may even be ruptured, but the united protoplasts become surrounded by a thick wall of two layers, the outer being warted and black (Fig. 3.10).

In most organisms fusing gametes are uninucleate and naked (i.e. without walls). In *Mucor* it is not easy to decide what should be called a gamete: whether the whole multinucleate contents of the gametangium, or an individual nucleus. Again, although the black, thick-walled structure resulting from sexual fusion is usually termed a zygospore, it might be more correct to refer to it as a zygosporangium containing a closely-fitting zygospore.

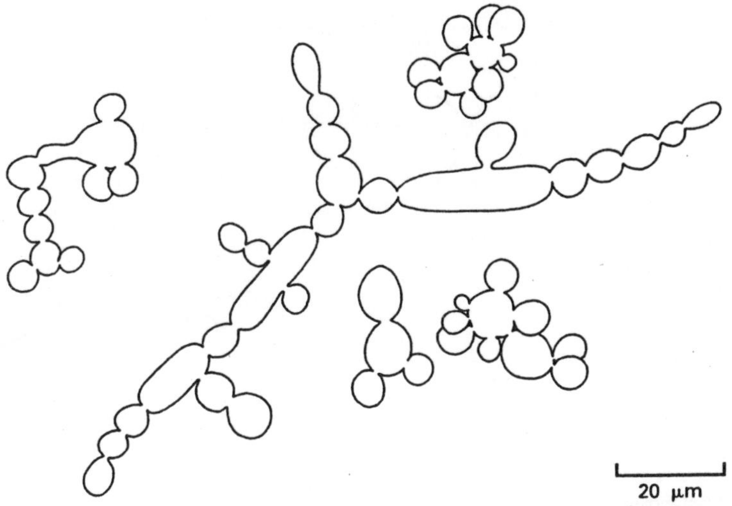

Fig. 3.9. *Mucor racemosus.* Yeast-like budding in liquid culture rich in sugar.

The sexual stage in *Phycomyces* is, in general, like that of *Mucor*, but the zygospore is surrounded by antler-like, black appendages, developed from the suspensors, which make it easily visible to the unaided eye. Further, apart from the sporangiophores, aerial hyphae are few. Thus when 'plus' and 'minus' colonies meet, a striking black line of zygospores is produced.

In *Phycomyces*, as colonies of opposite sign approach each other, certain submerged hyphae swell and on meeting grasp each other to form a knot-like structure within the agar. Still adhering they grow upwards together, but on emerging above the surface their tips arch apart as the progametangia. Gametangia are then cut off and fusion occurs as in *Mucor*, but as the zygospore develops the branched antler-like appendages grow from the suspensors (Fig. 3.11).

Not all mucoraceous fungi are heterothallic. Some are homothallic as in *Rhizopus sexualis* in which the zygospore is formed between neighbouring branches of the aerial zygophore (Fig. 3.12; Plate 1b). Further, although in most Mucorales there is no consistent difference in size in the fusing gametangia, in a few the zygospore is formed by the union of a larger with a much smaller gametangium. Where this happens the fungus is always homothallic. This condition occurs in *Zygorhynchus*, a common genus in soil (Fig. 3.13).

The zygospore is not a dispersive spore. Rather it is a resting-spore in the sense that it cannot germinate immediately, but must first undergo a period of dormancy. This seems to vary with the species from a few weeks to several months. However, such little evidence as there is suggests that zygospores are not particularly long-lived, and after a year or so most seem to lose their capacity to germinate.

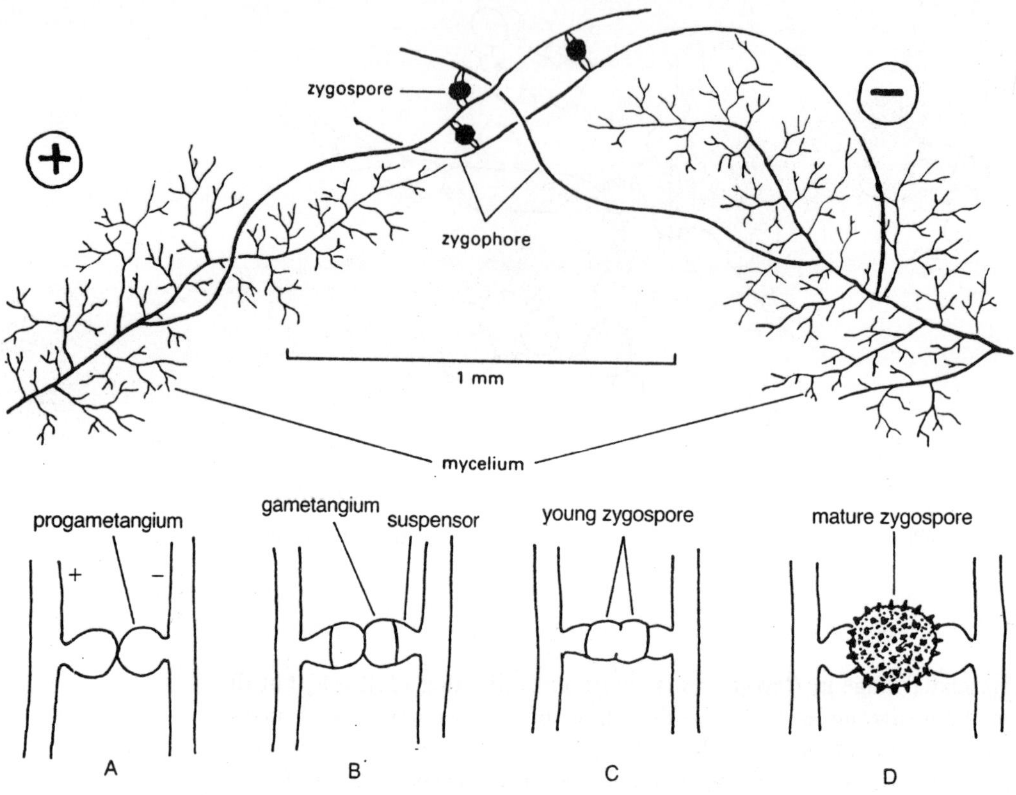

Fig. 3.10. *Mucor hiemalis*. Above, region where 'plus' and 'minus' colonies are approaching: zygospores have been produced between the aerial zygophores. Below (A–D), stages in zygospore formation: A, progametangia in contact; B, gametangia delimited; C, gametangia fusing; D, mature zygospores flanked by suspensors.

At germination a germ tube develops from the zygospore and normally grows into a sporangiophore bearing a sporangium of the normal type, but known as a germ-sporangium (Fig. 3.14). In *Mucor*, analysis of the spores from this sporangium shows that they are all 'plus' or all 'minus'. We have seen that in the formation of the zygospore the haploid 'plus' and 'minus' nuclei probably fuse pairwise giving a number of diploid nuclei, although some nuclei may remain unpaired. It is generally agreed that meiosis occurs at some stage during zygospore germination. In an attempt to explain the fact that all spores from a germ-sporangium are of the same sign, it has been suggested that in *Mucor* meiosis is confined to a single surviving nucleus, the others degenerating. It is further proposed that following meiosis only one of the four haploid nuclei survives (Fig. 3.15). It must be emphasized that there is no direct confirmation of this from cytological studies. Technically it is extremely difficult to follow the cytological story because the nuclei are so small.

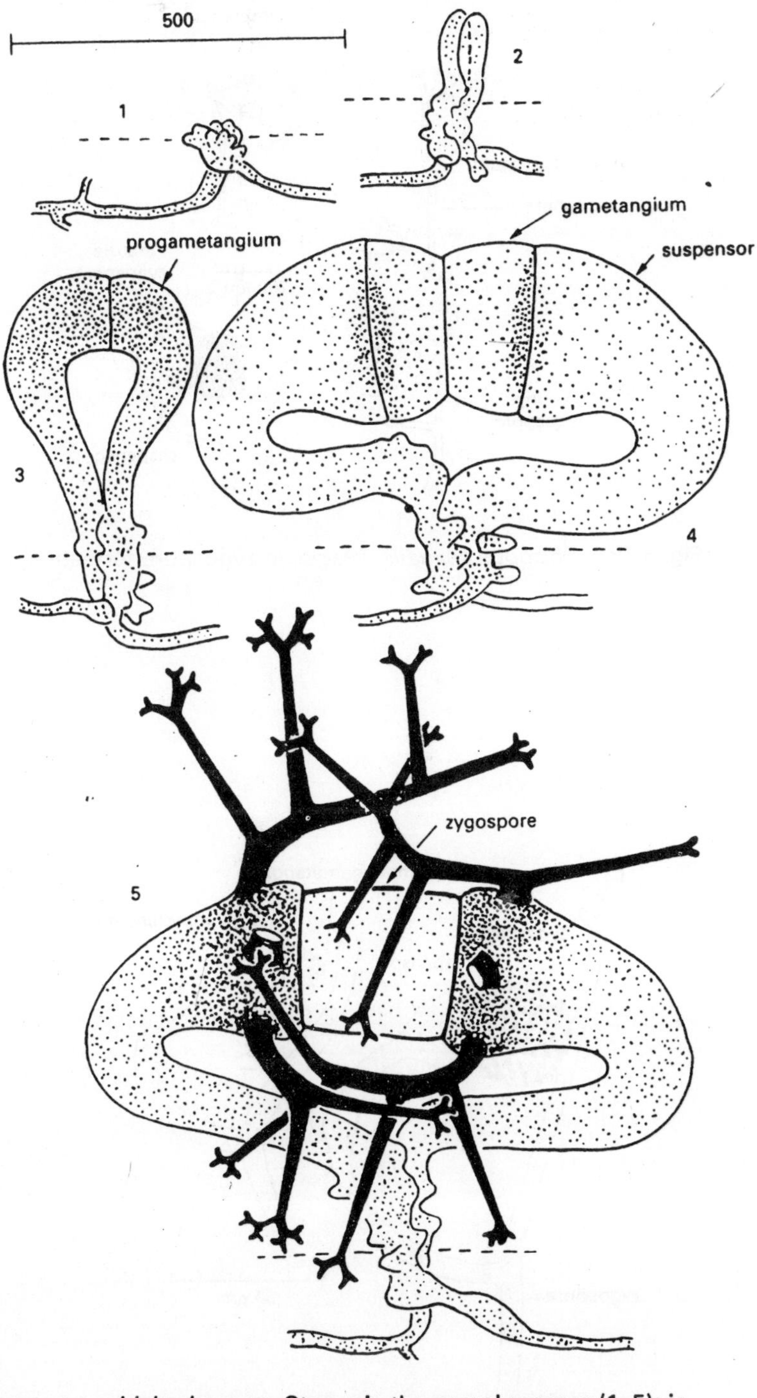

Fig. 3.11. *Phycomyces blakesleeanus*. Stages in the sexual process (1–5): in 5 two of the appendages of the suspensors on the near side are shown cut short, and those on the far side are omitted. Dashed lines indicate the surface of the agar.

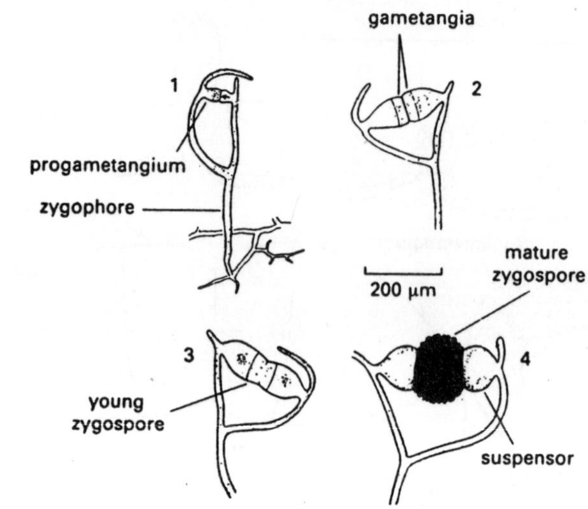

Fig. 3.12. *Rhizopus sexualis.* Stages in zygospore formation.

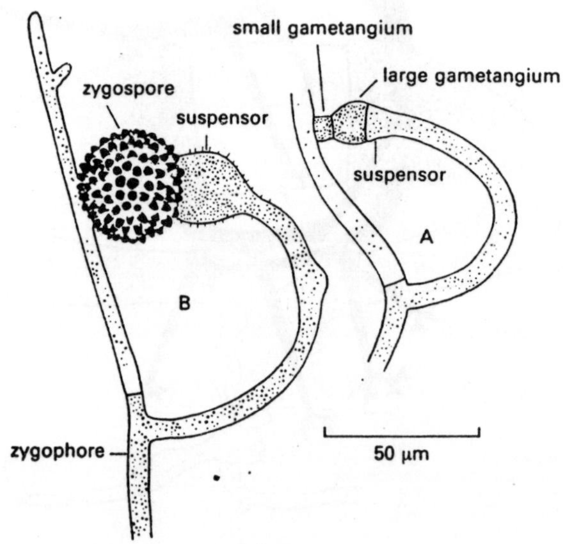

Fig. 3.13. *Zygorhynchus.* A, two gametangia of unequal size about to fuse. B, later stage with mature zygospore.

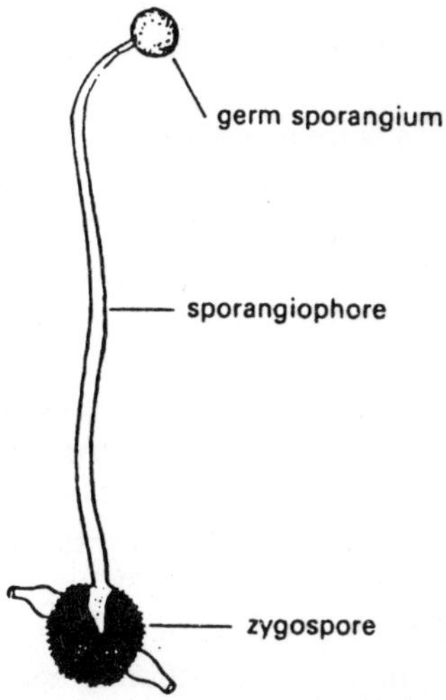

Fig. 3.14. *Mucor* sp. Zygospore, considerably magnified, with its suspensors attached, germinating to give a germ-sporangium. (After Brefeld)

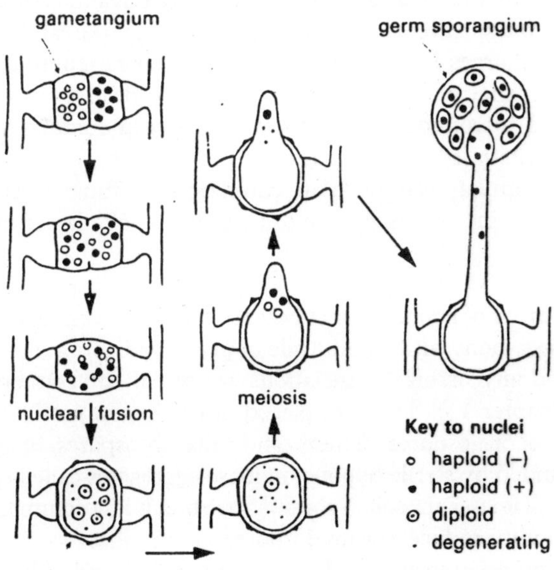

Fig. 3.15. *Mucor.* Diagram of suggested nuclear behaviour in zygospore formation and its germination to give a germ-sporangium. If, following meiosis, the surviving nucleus is 'minus', the spores in the sporangium would be of that sign and not 'plus' as in the diagram.

In *Phycomyces,* in contrast to *Mucor,* the germ-sporangium contains both 'plus' and 'minus' spores. In this fungus it would thus appear that all the products of meiosis of one or more diploid nuclei contribute to the protoplasm of the germ tube from which the sporangiophore with its germ-sporangium develops.

Although Mucorales is quite a small order of the fungi, it has been given rather extensive treatment here. This is partly because so many of the fundamental aspects of fungi are so well illustrated by these fungi, and partly because they are ideal fungi to study because they grow so readily in culture.

Entomophthorales

Entomophthorales is another small but interesting order of Zygomycetes, its members mostly being parasites of insects. Asexual reproduction is by conidia which are usually violently discharged. Some species form resting-spores which may be produced as the result of a sexual process not unlike zygospore formation in Mucorales.

A well-known species is *Entomophthora muscae* (Fig. 3.16) which attacks house-flies and hoverflies in late summer and autumn. The discharged conidium adheres to the fly and a germ tube penetrates its cuticle. As a result of infection the fly becomes plugged with the fungus, its abdomen in particular being considerably inflated. In its death agonies the fly may cling to a blade of grass or, quite often, to a window pane, by its sucker-like mouth-parts, and die in this attached condition. Shortly after death, conidiophores in great numbers burst through the thinner parts of the abdomen. Each conidiophore discharges its conidium, just as the sporangium is shot off in *Pilobolus*, to a distance of a centimetre or two. If the dead fly is on a window pane, it becomes surrounded by a halo of discharged conidia. The fungus attack on flies sometimes reaches epidemic proportions.

E. muscae is a familiar example of an entomogenous fungus. Although not numerous, fungi living on insects are to be found in all classes of fungi.

Endogonales

The Endogonales are of major interest because they form vesicular-arbuscular mycorrhizal associations with a multitude of plant types, from mosses to seed plants. Such endomycorrhizal associations are by far the commonest of all mycorrhizas (Chapter 11). They are placed in the Zygomycotina because of the production of zygospores. These, and chlamydospores, may occur in clusters, surrounded by sterile hyphae, in loose aggregates called sporocarps, some 20 mm in diameter, in soil. Sporangia are much less common and most British species only produce chlamydospores.

Chlamydospores arise as balloon-like swellings at the end of hyphae. They become thick-walled, pigmented, yellow to dark brown or black, and are up to 250 μm in diameter (Plate 4c).

MASTIGOMYCOTINA

The Mastgoimycotina are zoosporic fungi and zoospore structure is an important feature in their classification. A zoospore is a uninucleate spore without a cell wall which can only exist in water where it swims by means of

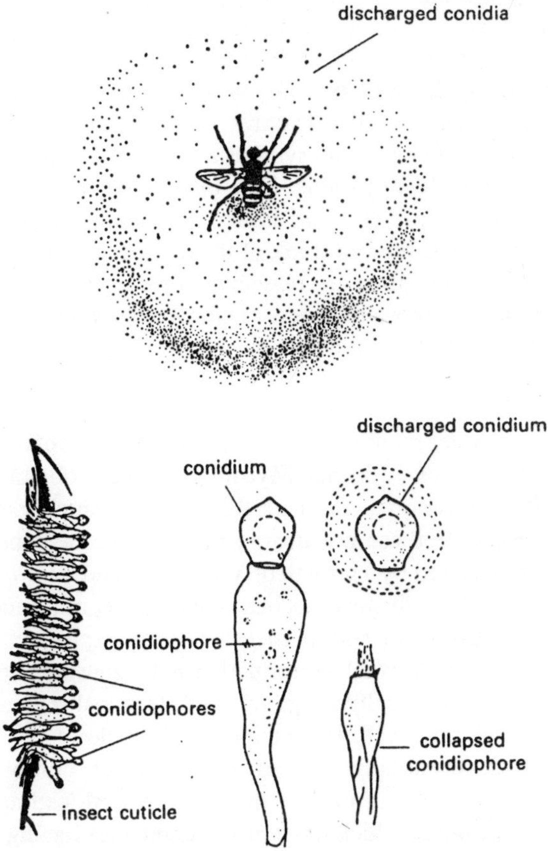

Fig. 3.16. *Entomophthora muscae.* Above, dead fly attached to glass by its mouth-parts and surrounded by a halo of discharged conidia: note the three white bands of conidiophores on its abdomen. Below left, part of a section through the abdomen of a dead fly showing conidiophores projecting through the integument of the insect (× 80). Below right, single mature conidiophore; conidiophore after discharge; and discharged conidium surrounded by conidiophore sap (× 320).

one or two flagellae. Many authorities now place them in the Protoctista rather than the fungi.

Oomycetes

The characteristic feature of Oomycetes is oogamous sexual reproduction with one or several eggs being formed in a spherical oogonium. The vegetative mycelium is non-septate and its nuclei are diploid, a most unusual situation in fungi. Meiosis occurs in the developing oogonium and antheridium to give haploid sexual nuclei. In Oomycetes the zoospore is always biflagellate, one flagellum being of the tinsel type which projects forward during swimming, and the other of the whip-lash type trailing behind (Fig. 3.17). The distinctions between these two kinds of flagellum, clearly seen only under the electron-microscope, is of considerable taxonomic importance. Another curious feature of members of this class, already mentioned, is that the microfibrils of the hyphal wall are composed of a form of cellulose, not chitin. Oomycetes have other peculiar features. Unlike true fungi, they do not translocate trehalose. They cannot synthesize, and thus require, sterols. They have a well-defined Golgi apparatus. Like plants, they synthesize the amino-acid lysine via diaminopimelic acid rather than aminoadipic acid as in the fungi. Even mycologists tend to think of them as occupying a rather isolated position in the fungi but they are reluctant to part with them.

Only the two most important orders of Oomycetes will be considered here: Saprolegniales and Peronosporales.

Saprolegniales

Saprolegniales are water moulds often to be seen as a fringe of hyphae around a dead fish floating in a pond or canal. Mycologists usually obtain specimens for study by a baiting technique. Autoclaved house-flies, ants' eggs or oily seeds form suitable baits. If these are floated in a sample of pond water, they normally become covered by a growth of *Saprolegnia* and *Achlya* in a few days.

Saprolegnia (Fig. 3.17) forms branched, non-septate, multinucleate hyphae in the nutrient substratum and branches bearing reproductive structures project into the water. Asexual reproduction is by zoospores produced in a zoosporangium. This is usually a cigar-shaped, terminal segment of a hypha delimited by a basal cross-wall. It contains a multinucleate mass of colourless protoplasm which eventually cleaves into a large number of naked, uninucleate zoospores. When mature they can be seen jostling one another within the zoosporangium which has a protuberant apex having a particularly thin wall. This suddenly breaks down and the zoospores escape. Most come out with a rush, as if discharged by pressure within the zoosporangium, but the remainder escape more slowly. Each is pear-shaped, with the two flagella at the pointed end. The zoospores escape blunt-end foremost. This is most clearly seen in the residual zoospores which emerge after the first mad rush. Once outside, the direction of the zoospore reverses and it swims pointed-end foremost propelled by the flagella; the tinsel flagellum directed forwards, the whip-lash one trailing.

After a period of motility, lasting only a few minutes, the zoospore withdraws its flagella and surrounds itself with a thin wall. It is said to encyst. This state is maintained for a few hours. Then the zoospore emerges, like a chicken from an egg, leaving the empty cyst behind. On emergence, however, the zoospore is in a different form, being rather bean-shaped and with the flagella laterally inserted. This zoospore swims about for up to a few hours and then again acquires a wall to become a secondary cyst. There may be a repeat of this performance, but eventually the secondary cyst, like a normal fungal spore, germinates by a germ tube to establish a mycelium.

Diplanetism, this curious two-stage behaviour of the zoospore, is a striking feature of a *Saprolegnia* and some of its relatives. Its biological significance, if any, is obscure. In *Saprolegnia* the first period of diplanetism is relatively long.

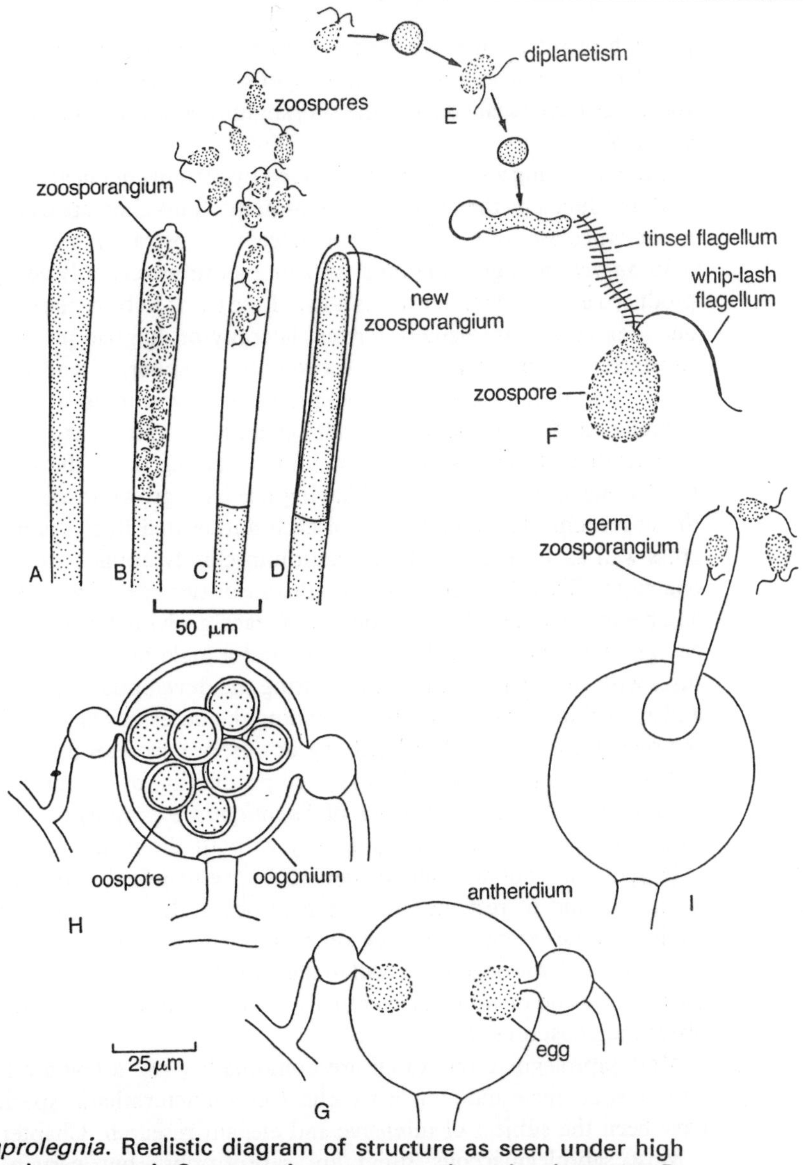

Fig. 3.17. *Saprolegnia*. Realistic diagram of structure as seen under high power of the microscope. A–C, stages in zoosporangium development. D, new sporangium produced by internal proliferation. E, stages in diplanetism. F, zoospore very highly magnified. G,H, stages in the sexual process leading to oospore production. I, germination of oospore within the oogonium; one has produced a germ-sporangium.

In *Achlya*, however, it is extremely short and the zoospores encyst immediately on escape forming a characteristic ball of primary cysts capping the empty zoosporangium.

A second feature serves to separate the two large genera *Saprolegnia* and *Achlya*. In *Saprolegnia*, after the zoospores have escaped, the cross-wall at the base of the empty sporangium bulges and forms a new hyphal apex which grows forward to produce a new sporangium within the old. This internal proliferation contrasts with what happens in *Achlya*. In that genus, after the zoospores have escaped from the sporangium, a new one grows up at the side of the old.

The swimming zoospores are attracted by certain chemicals, notably amino-acids produced by the breakdown of proteins. This chemotaxis no doubt helps in the 'selection' of a suitable proteinaceous substratum such as a dead fish.

In *Saprolegnia*, growing on a suitable bait in water, only zoosporangia are produced at first, but later the asexual phase tends to be replaced by the sexual one. Sex organs are usually formed laterally on hyphae projected from the substratum. The female organ (oogonium) is spherical and delimited by a basal cross-wall (Plate 1a). The protoplasm within is at first uniformly granular and multinucleate, but eventually, after many nuclei have degenerated, it cleaves into a few uninucleate eggs. In the meantime from nearby one or more fine antheridial hyphae have grown towards and invested the oogonium. The end section of each antheridial hypha is delimited by a cross-wall as a multinucleate antheridium closely applied to the wall of the oogonium. Then from the antheridium one or two specialized hyphae develop and penetrate the wall of the oogonium, each growing towards the nucleus of an egg. The hyphal tip then ruptures and a male nucleus is liberated. This fuses with the nucleus of the egg. During the development of both oogonium and antheridium meiosis occurs providing the necessary haploid condition for the sexual nuclei. The fertilized egg or oospore secretes a smooth, thickish wall and is a resting spore. This sexual production of oospores is the typical behaviour, but in many species of *Saprolegnia* a proportion of the oogonia contain oospores formed apogamously without the mediation of antheridia.

Oospores are not normally dispersed, but remain within the old oogonium. After a period of rest germination may occur. In this process the oospore swells to some extent, its wall becomes thin, and a germ tube is produced that grows through the oogonium wall to the outside. There a club-shaped hypha is formed, the end of which is eventually delimited as a small germ-sporangium liberating zoospores (Fig. 3.17).

Most saprolegniaceous fungi are homothallic, but a few are heterothallic with separate male and female mycelia. Certain heterothallic species of *Achlya* have been the subject of intensive and elegant research (Chapter 8).

Most saprolegniaceous fungi are saprotrophs, but certain species of *Saprolegnia*, especially *S. diclina*, are often responsible for the death of salmonid fish. Although only the superficial tissues of the host become infected, the osmoregulatory mechanism of the fish is upset and recovery rarely occurs.

Peronosporales

Peronosporales is an order of Oomycetes the members of which range from certain truly aquatic saprotrophic and facultative necrotrophic species of *Pythium* and *Phytophthora* to the downy mildews (Peronosporaceae), terrestrial fungi that are obligate biotrophic parasites of the shoots of higher plants.

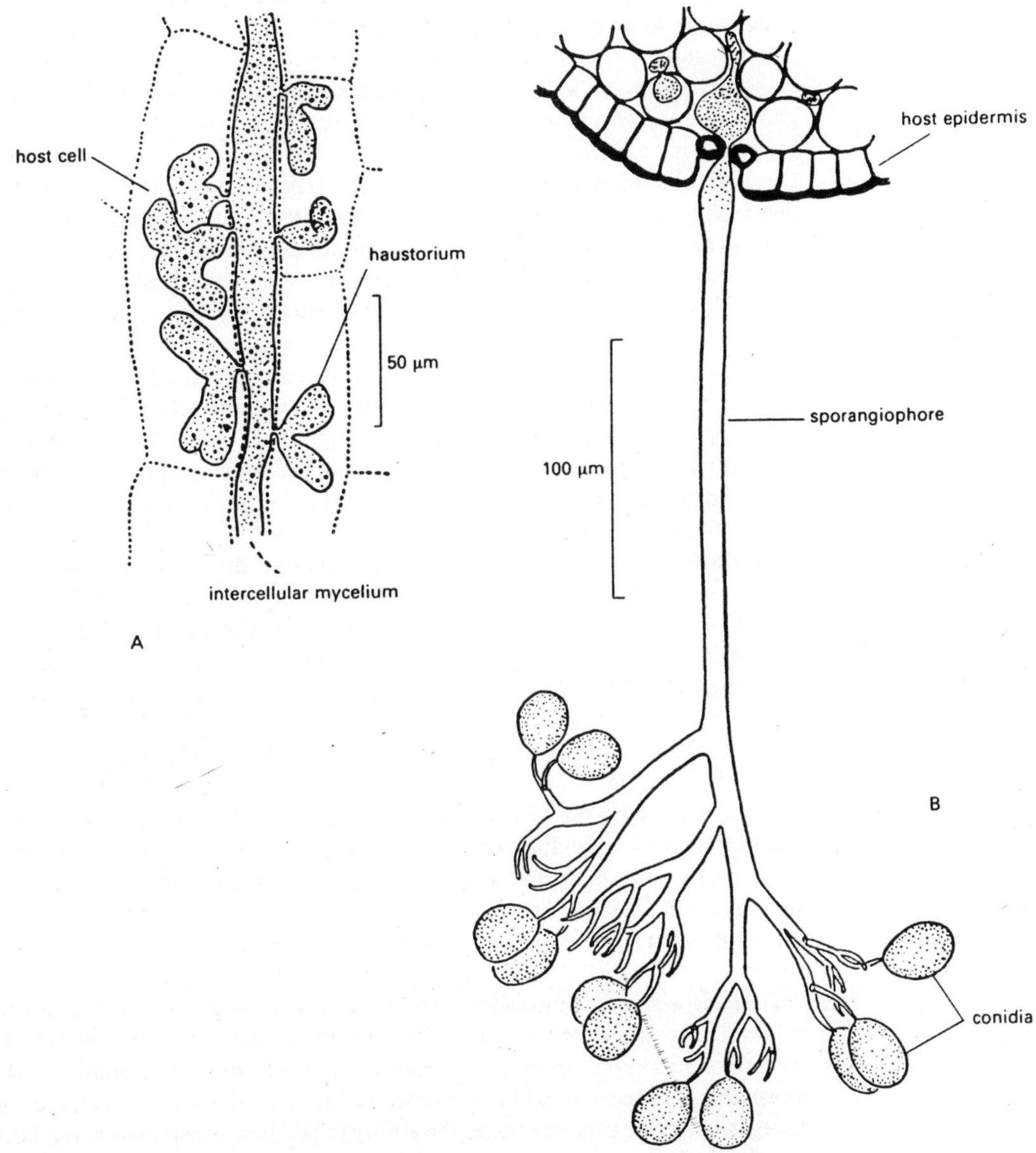

Fig. 3.18. *Peronospora parasitica.* A, mycelium as seen in longitudinal section of the pith of a diseased wallflower (*Cheiranthus*); living host cells shown with dotted outline. B, sporangiophore emerging through the stoma of a host plant.

Peronospora parasitica (Fig. 3.18) may be taken as a common example of this order. It is an obligate parasite of wallflower (*Cheiranthus*) and shepherd's purse (*Capsella bursa-pastoris*), an abundant garden weed. In a diseased wallflower a group of terminal leaves and the associated sappy stem appear somewhat reduced and distorted, and have a white, mildewed appearance. This is due to a covering of sporangiophores.

The young leaves of a healthy plant are infected by airborne spores. The spore germinates in a drop of dew or rain on the leaf surface, and puts out a germ tube which grows into the leaf by way of a stoma. Once inside, the fungus forms a branched, non-septate mycelium in the intercellular spaces of the host obtaining nourishment through relatively large forked haustoria developed within the living cells. The parasitized cells are not killed; the success of the parasite depends on the continued life of the penetrated host cells. After a few days of vegetative activity within the host, reproductive structures are formed. These are branched sporangiophores. Each emerges through a stoma and has an unbranched lower region and a branched upper one with fine ends each bearing an oval spore, which is regarded as a modified sporangium. The spores in *Peronospora* are very finely attached. Just how they are liberated is not quite clear. In some species of the genus they appear to be violently discharged as the result of twirling of the sporangiophore axis during sudden drying. It has also been suggested that static electric charges play a part in their liberation.

In Peronosporaceae, the family of the downy mildews, the individual genera are characterized by the form of the branched sporangiophore. In all genera, except *Peronospora,* the spore in the infection-drop on the leaf can behave in one of two ways. It can germinate directly, thus resembling a conidium; or, behaving as a sporangium, it can first liberate zoospores each of which, after encystment, gives rise to a germ tube. Whether a germ tube comes directly from the sporangium, or indirectly from an encysted zoospore, entry of the fungus to the host is by way of a stoma. Where indirect germination occurs the protoplasm of the spore (sporangium) cleaves into a small number of individual segments which become the zoospores. The apex of the sporangium then dissolves and the bean-shaped zoospores, like those of *Saprolegnia* in the second stage of diplanetism, swim in the infection drop. It is not clear what factors determine whether direct or indirect germination occurs.

There is a strong probability that these terrestrial fungi causing downy mildew in higher plants are derived from ancestors that were water moulds. In this transmigration from the aquatic to the land environment, the zoosporangium has tended to be converted into a conidium. *Peronospora* can be regarded as the ultimate stage, the ability to produce zoospores having been completely lost.

As well as reproducing asexually, *Peronospora parasitica* has a sexual stage. This occurs in the deeper parenchymatous tissues of infected wallflower shoots. Oogonia and antheridia are formed on the mycelium in the larger intercellular spaces of the host plant. The sexual process is similar to that of *Saprolegnia,* except that in Peronosporales only one egg is found in the oogonium. After fertilization the egg is converted into a thick-walled oospore, a resting-spore. It is set free only when the diseased tissue dies and rots.

In any fungus that causes plant disease, the question of hibernation is a significant one. In many pathogens the unfavourable period is endured as a sexually produced resting-spore. In the spring this may germinate and, if it happens to be in contact with the appropriate host, initiate infection.

Although *Peronospora parasitica* is not an important pathogen, other members of Peronosporaceae may cause serious disease. Downy mildew of vine caused by *Plasmopara viticola* is one of the most destructive diseases of vineyards, *Pseudoperonospora humuli* is an important pathogen of hops, *Peronospora destructor* causes heavy loss in areas of large-scale onion cultivation, and *Bremia lactucae* may cause serious trouble as a parasite of lettuce.

Pythiaceae is another important family of Peronosporales. It contains the two large genera *Pythium* and *Phytophthora*. In the genus *Phytophthora*, with about 40 species, there are pathogens of great economic importance. Some species of the genus are saprotrophic water-moulds growing on decaying leaves submerged in streams, while others are specialized parasites of higher plants closely approaching the downy mildews.

One such species *Phytophthora infestans* (Fig. 3.19), the cause of potato blight, deserves special attention not only because of its economic importance, but also because of its significance in the development of the science of plant pathology. In its mode of attacking its host and in its structure this pathogen is very similar to *Peronospora*. However, after the sporangiophores are produced, the host tissue dies. Further, the fungus can be grown in culture on agar, although in nature it is found only in the parasitic condition.

Blight normally strikes the potato crop during August in warm, damp weather. At an early stage in the disease individual leaflets have small, dead areas (Fig. 3.19). On the underside these can be seen to be fringed with an infected but still living zone in which white sporangiophores are apparent. The dead areas rapidly enlarge and the whole leaf is killed. Sporangia are blown away and cause fresh infection in nearby plants. Others may fall to the ground, where they are washed into the soil and infect the tubers. In a surprisingly short time, under suitable weather conditions, the potato haulms are reduced to a black putrid mass; for once the shoots are killed by the fungus, saprotrophic bacteria infest the dead tissue and cause putreation.

P. infestans occupies a special place in the history of plant disease and in history generally. It was this fungus, introduced into Europe from the New World in the late thirties of last century, that destroyed the potato crop in Ireland in 1845 and in subsequent years causing the Irish famine. The investigations of Berkeley in Britain and of de Bary in Germany led to the then surprising conclusion that the microscopic fungus associated with the rotting potato plants was the causal organism of the disease and not merely a by-product of decay. As the nineteenth century progressed, more and more fungi were found to be concerned in plant disease, and it became apparent that fungi are the most important pathogens with which the plant pathologist has to contend.

P. infestans continued to ravage the potato crop whenever weather conditions were suitable, until the discovery of spraying with a fungicide as a means of combating the disease. This development was due to the French chemist Millardet who, in 1882, observed that vines sprayed with a mixture of

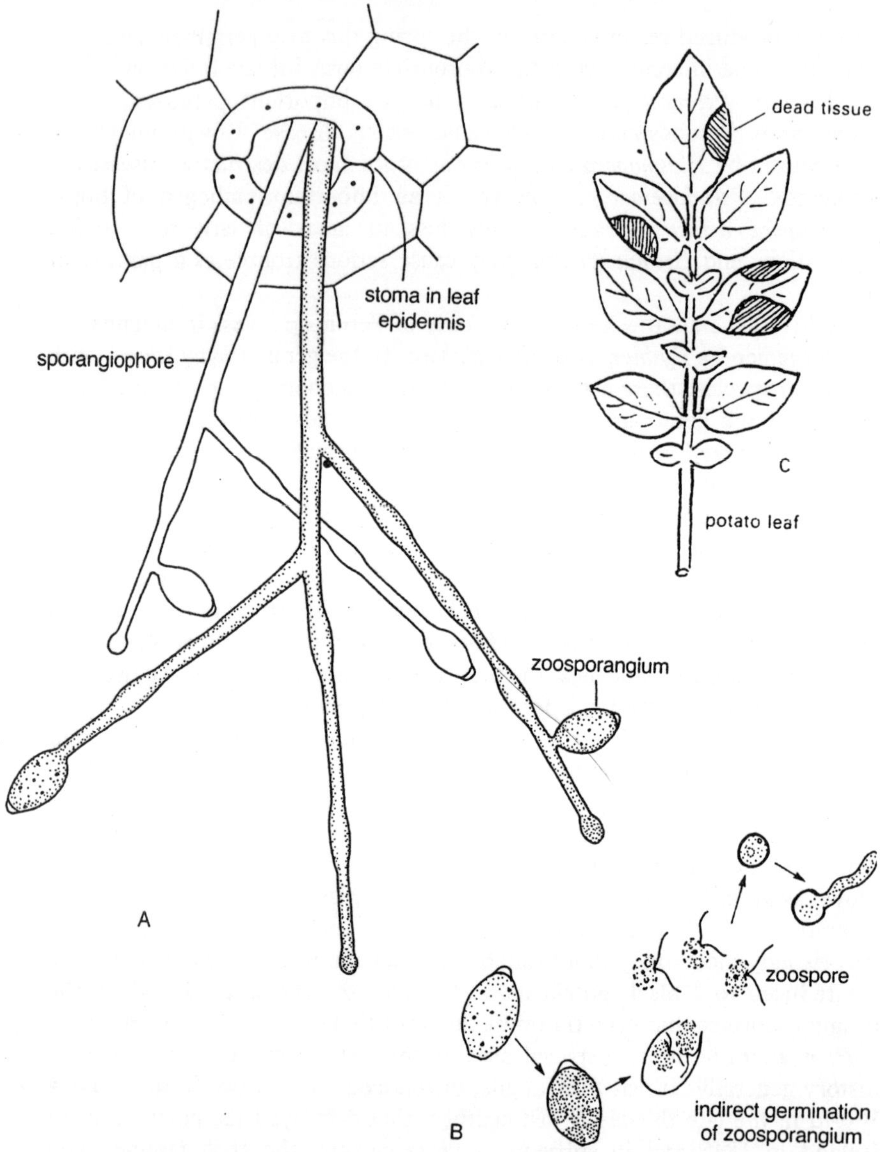

Fig. 3.19. *Phytophthora infestans*. A, two sporangiophores grown out through the stoma (shown in surface view) of a potato leaf. B, stages (linked by arrows) in the indirect germination of a spore (zoosporangium); the encysted zoospore finally germinates to give a germ tube. C, early stage in the infection of a potato leaf.

copper sulphate and lime to discourage pilfering of the grapes were free from the mildew, *Plasmopara viticola*. It was realized that a spray, effective against that disease, might control the rather similar potato-blight fungus. The spray developed by Millardet was called Bordeaux Mixture. It consists of a fine aqueous suspension of cupric hydroxide formed by mixing copper sulphate and lime water in suitable proportions. When this is sprayed on potato foliage so as to wet both upper and lower leaf surfaces, before the sporangia of *P. infestans* become prevalent in the air, good control is obtained. Cupric hydroxide is insoluble in water, but is very slightly soluble in water containing germinating sporangia, apparently because of the production of organic acids by the fungus. Thus as the sporangia germinate in the infection drop, enough copper comes into solution to kill the germ tubes. In spraying, timing is vital. If it is done too early, the fungicide may be washed off by rain, and new unprotected foliage may be produced. If spraying is too long delayed, infection may have already occurred. In a major potato-producing area the authorities in late summer issue warnings to spray when meteorological conditions indicate that epidemic spread of the fungus is likely. In addition to control by spraying, plant breeders have put much effort into developing varieties resistant to blight.

We have seen that *Peronospora parasitica* overwinters by resting-spores produced sexually. However, although in *Phytophthora infestans* these can be induced to form under conditions of pure culture in the laboratory, they are very rarely seen under field conditions. The potato blight fungus normally passes the winter as a dormant mycelium within the tissues of lightly infected tubers.

Chytridiomycetes

Judged by number of species, Chytridiomycetes is the largest of the classes of the Mastigomycotina. The characteristic feature of the class is asexual reproduction by zoospores each with a single, posterior flagellum of the whip-lash type.

Chytridiales

Chytridiales or chytrids are mostly aquatic fungi and many are so small as to be recognizable only under the high power of the microscope. They are especially noteworthy as parasites of both filamentous and unicellular algae. Most species of algae contributing to the phytoplankton of lakes (diatoms, desmids, peridinians, blue-green algae) have their specialized chytrid parasites. Chytrids also occur as saprotrophs on dead algae, on leaves of dead aquatic plants, and on various forms of chitin, cellulose and keratin in water and soil. Mycologists often obtain chytrids by 'baiting' a sample of pond water or a soil suspension. Baits commonly used are cellophane, insect wings, hair and pollen grains, particularly those of pine. A few chytrids parasitize higher plants.

A feature of most chytrids is that growth is limited. Usually the mature fungus consists merely of a zoosporangium and a rhizoidal system. The zoospore comes to rest on a suitable substratum and secretes a wall. It then puts out one or more fine rhizoids which branch to produce a system which is sometimes extensive but is always limited. These rhizoids contain cytoplasm

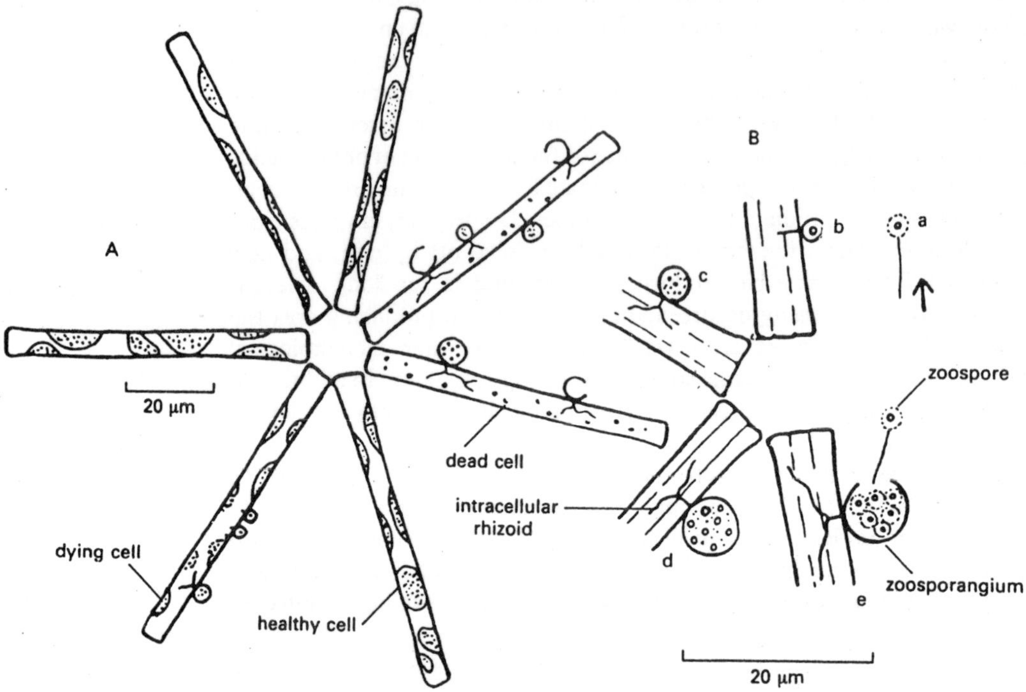

Fig. 3.20. *Rhizophydium planktonicum* on the diatom *Asterionella*. A, colony of seven diatoms; four living, one dying with young stages of the parasite, and two dead with later stages. B, the central part of a diatom colony showing diagrammatically the asexual cycle (a–e).

but no nuclei, and are thus not the equivalent of ordinary fungal hyphae. The original body of the zoospore enlarges into a zoosporangium which is often spherical. In this the zoospores are delimited. They eventually emerge through specialized holes that develop in the wall. Escape of the zoospores ends the life of the fungus. In a few chytrids there is no rhizoidal system, the whole fungus being converted into a zoosporangium.

Many of the saprotrophic, monocentric chytrids have been grown in pure culture on agar, but this must be kept very moist. The culture consists not of a single mycelium, but of many separate individuals. As zoosporangia mature, the zoospores ooze out and give rise to new individuals just outside the old ones.

Many chytrids form resting-spores, often associated with a sexual process.

As a specific example of a chytrid, *Rhizophydium planktonicum* (Fig. 3.20) may be considered. This is a parasite of the diatom *Asterionella formosum* which frequently dominates the phytoplankton of large lakes. The population of this diatom can be considerably modified by an epidemic of the chytrid. The zoospore comes to rest on the diatom, acquires a wall and puts a fine haustorial rhizoid, which may branch, into the host cell which soon dies. The encysted zoospore then enlarges into a spherical zoosporangium in which a

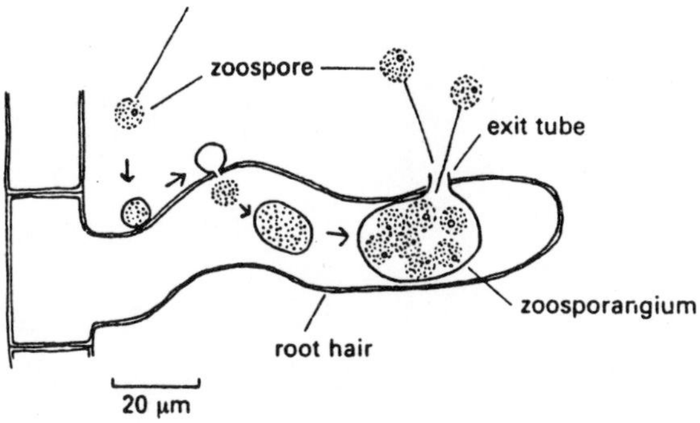

Fig. 3.21. *Olpidium brassicae.* Diagram of the asexual life-cycle on the root-hair of cabbage.

number of zoospores are delimited. An area of the wall then dissolves and the zoospores escape.

A rather different type is *Olpidium brassicae* (Fig. 3.21). It is a common parasite of cabbages grown in wet soil, although it does not cause serious disease. Cells of the piliferous layer of the roots are invaded. The zoospore settles on a root hair and injects its contents into the host cell. The chytrid protoplast then secretes a wall and enlarges into a zoosporangium; there is no rhizoidal system. At maturity the zoosporangium produces an exit tube which pierces the wall of the root hair. When the tip of this tube dissolves, the zoospores swim out into the soil water. Later two zoospores may act as gametes and fuse outside the host. The zygote infects producing a characteristic thick-walled resting sporangium. As the root decays these are released into the soil. After several months, following a nuclear fusion and a meiosis, germination occurs to produce haploid zoospores.

There are numerous other equally diverse classes in both the Zygomycotina and the Mastigomycotina which are not considered in this book. There is little evidence of how, if at all, the different classes may be related to one another phylogenetically, or indeed, to the 'higher fungi'. All these fungi seem to have in common is a relative simplicity of structure.

Ascomycotina | 4

Ascomycotina is the largest division of fungi. If lichens are included, there are over 42 000 species. In spite of a considerable range of form, the division appears to be a natural one. Its members are characterized by having asci. The ascus is a special type of sporangium which usually contains eight ascospores. Development is highly distinctive. The young ascus has two haploid nuclei. These fuse to form a diploid one which immediately undergoes meiosis so that four haploid nuclei are formed. Usually each of these then divides mitotically. Around the individual nuclei ascospores are organized, but some protoplasm (epiplasm) is left outside the eight ascospores (Fig. 4.1). In most Ascomycotina the ascus bursts at maturity, squirting its contained spores into the air to a distance varying with the species concerned from 2–400 mm (Fig. 4.2 and Plate 1d).

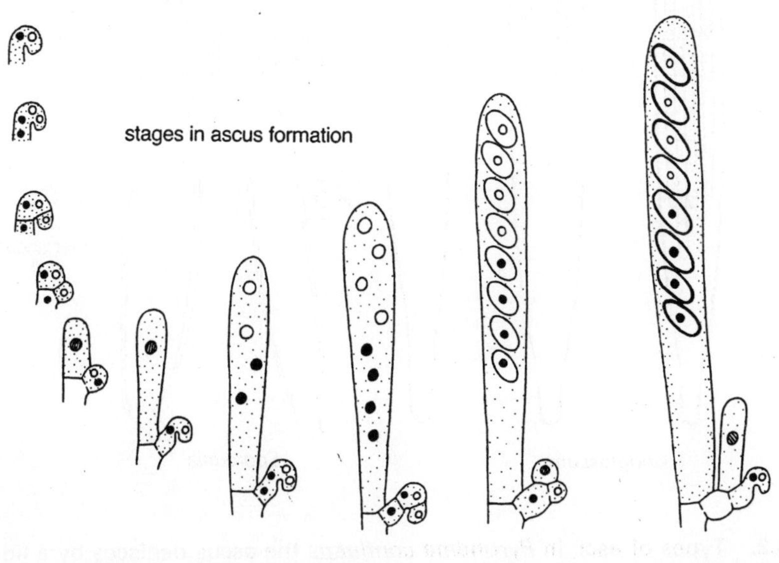

Fig. 4.1. Diagram of ascus development from the tip of an ascogenous hypha. Compatible haploid nuclei shown as black dots and white circles. Diploid nuclei shaded.

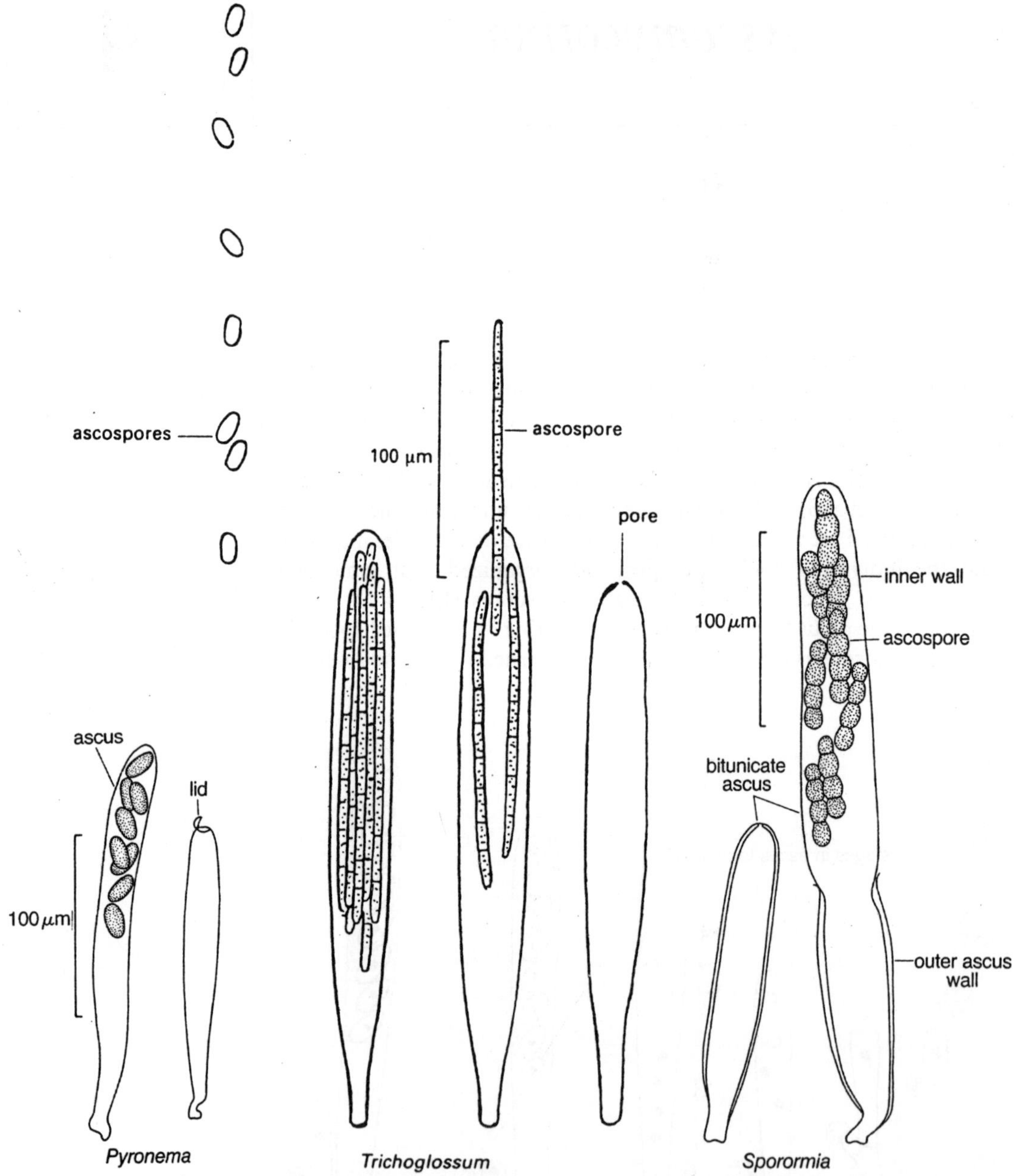

Fig. 4.2. Types of asci. In *Pyronema confluens* the ascus dehisces by a lid and the ascospores are discharged simultaneously. In *Trichoglossum hirsutum* the ascus dehisces by a pore and the ascospores are shot away in succession. In *Sporormia intermedia* the ascus is bitunicate; the inner wall finally bursts and the ascospores are discharged.

In the majority of Ascomycotina the asci are grouped together into complex ascocarps, although in a few 'lower' types they are produced singly.

The mycelium is branched and septate, the individual cells containing one to several nuclei. Each cross-wall is perforated by a minute central pore, usually hard to see under the microscope, through which mitochondria can easily pass; as can nuclei, but with more difficulty. As in septate fungi generally, vegetative fusions (anastomoses) frequently occur between nearby hyphae of the mycelium.

Although typically Ascomycotina reproduce by ascospores, a great many have an additional asexual means of reproduction, namely by conidia. In some the accessory spore-form has become dominant and the ascus stage is rarely formed. Purely conidial fungi, which are classified in Deuteromycotina (Fungi Imperfecti), probably represent, in the main, Ascomycotina which have lost the ascus stage completely or in which it has not yet been discovered.

In fungi the stage associated with nuclear fusion (karyogamy) and meiosis is now called the teleomorph, while the asexual (conidial) stage is known as the anamorph. Both together make up the holomorph or whole fungus. In a few Ascomycotina the holomorph embraces more than one anamorph.

The taxonomy of Ascomycotina is a tricky matter on which mycologists hold diverse views. However, the broad outline set out below would be acceptable to most. Ascomycotina may be grouped into a number of classes:

1 *Discomycetes* or cup-fungi in which the asci are packed side by side in an apothecium with the hymenium freely exposed.
2 *Pyrenomycetes* or flask-fungi where the asci are contained in a flask-like perithecium.
3 *Plectomycetes* in which the asci occur within a closed, spherical cleistothecium. Two very different orders are included: *Eurotiales*, an order of abundant saprotrophic moulds usually found as anamorphs (conidial state); and *Erysiphales* or powdery mildews, an important order of obligate plant parasites.
4 *Hemiascomycetes* in which the asci are not grouped in ascocarps. The most important family is *Saccharomycetaceae*, the yeasts.
5 *Loculoascomycetes* in which the asci occur in a pseudothecium which closely resembles a perithecium in form and function. However, it develops differently and the ascus has a double wall.

One other class, the Laboulbeniomycetes, must be mentioned, but will not be considered further in this book. In this class there are 130 genera, with 1500 species of minute fungi, only just visible with a hand lens, which are ectoparasites of insects. Members of this class are difficult to spot as they usually look like a regular part of the host. They cause little inconvenience to the infected insects.

DISCOMYCETES

Peziza may be considered as a typical member of Discomycetes. Species of this common genus are saprotrophs on rotten wood and they also occur in

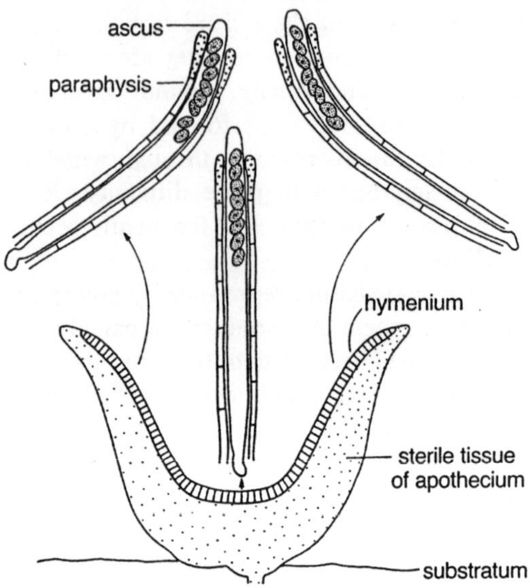

Fig. 4.3. *Peziza (Aleuria) vesiculosa*. Apothecium in vertical section. Hymenium shown hatched. An ascus and two paraphyses are shown taken from three indicated positions in the hymenium. (Apothecium × 2, ascus × 80. Based on figures by Buller)

forest soils rich in organic matter. In *P. vesiculosa* (Fig. 4.3) the branched, septate mycelium is in the soil while the ascocarp, an apothecium, is formed above ground. An ascocarp is an apparatus concerned with the production and liberation of spores. This is commonly short-lived. The perennial feeding mycelium remains out of sight in the nutrient substratum.

The apothecium of *P. vesiculosa* is cup-shaped, one to several centimetres across and of a pale fawn colour. Another common cup-fungus, *Aleuria aurantia*, is brilliant red–orange and at a distance looks like a piece of orange peel.

A vertical section of the apothecium (Fig. 4.3) reveals that it is lined by a hymenium consisting of long asci in all stages of development interspersed with filamentous packing hyphae (paraphyses). The general tissue of the ascocarp is pseudo-parenchymatous. Although ascocarps may be fairly substantial structures, they are never truly parenchymatous, but consist of densely interwoven hyphae.

The asci and paraphyses are positively phototropic so that their ends point towards the incident light. With a cup-shaped apothecium with light from above, the asci at the base are straight, but those on the side walls are curved (Fig. 4.3). Each mature ascus is a single cell surrounded by a cell wall considerably stretched by the hydrostatic pressure within. Lining the wall is a very thin layer of cytoplasm around a large vacuole of cell sap in which the oval, unicellular ascospores are suspended near the apical end (Plate 1d). Finally the turgid ascus bursts. A little lid hinges backwards, the ascus wall

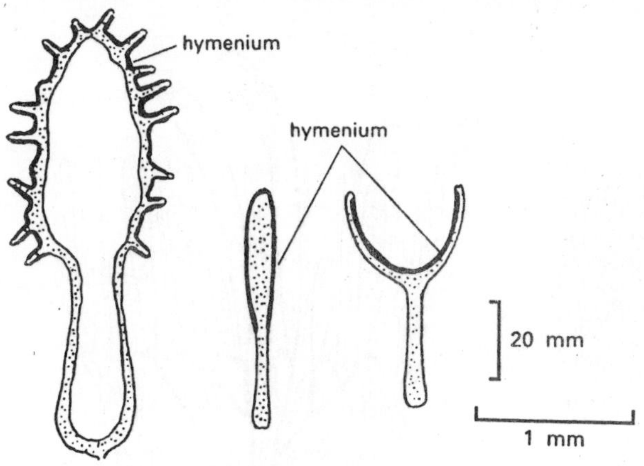

Fig. 4.4. Apothecia of some large Discomycetes shown in vertical section: *Morchella, Geoglossum (Trichoglossum), Macroscyphus*

contracts and the spores are squirted into the air to a distance of 20–30 mm. This is illustrated in *Pyronema* (Fig. 4.2).

The phototropism of the ascus seems to have a biological advantage. If instead of curving in response to light, the asci were all straight and set at right-angles to the lining of the apothecium, a proportion of the ascospores discharged from the more upright walls of the cup would be wasted by being shot onto the opposite wall.

In cup-fungi generally 'puffing' occurs. During a period of quiescence thousands of asci reach a superstretched condition of unstable equilibrium. If an apothecium in this state is touched or even breathed upon, these burst simultaneously, throwing into the air a cloud of ascospores. Although the individual spores are too small to be seen by the unaided eye, in the mass, like motes in a sunbeam, they form a visible cloud.

Most Discomycetes are small, only a few millimetres in diameter, and the apothecium is in the form of a cup or saucer. However, other shapes occur (Fig. 4.4), the essential feature being that the hymenium is freely exposed. In *Macroscyphus* the cup-like apothecium is stalked. In *Trichoglossum* it is club-shaped. In *Morchella*, the largest of the Discomycetes and a much-prized edible fungus, the ascocarp has a sterile stalk bearing a fertile head the surface of which is highly convoluted to form broad depressions lined by hymenium.

The development of an apothecium can be illustrated by reference to *Ascobolus stercorarius* in which the process has been carefully studied. In the genus *Ascobolus* the ripe ascospores are purple, and the phototropic asci project considerably beyond the general level of the paraphyses (Fig. 4.5). In the coprophilous succession, species of *Ascobolus* regularly follow the *Pilobolus* phase. About a week after placing fresh horse-dung balls under a transparent cover with adequate aeration, hundreds of apothecia usually develop. If the cover is only 5–10 cm above the dung, great numbers of purple spores are shot on to it.

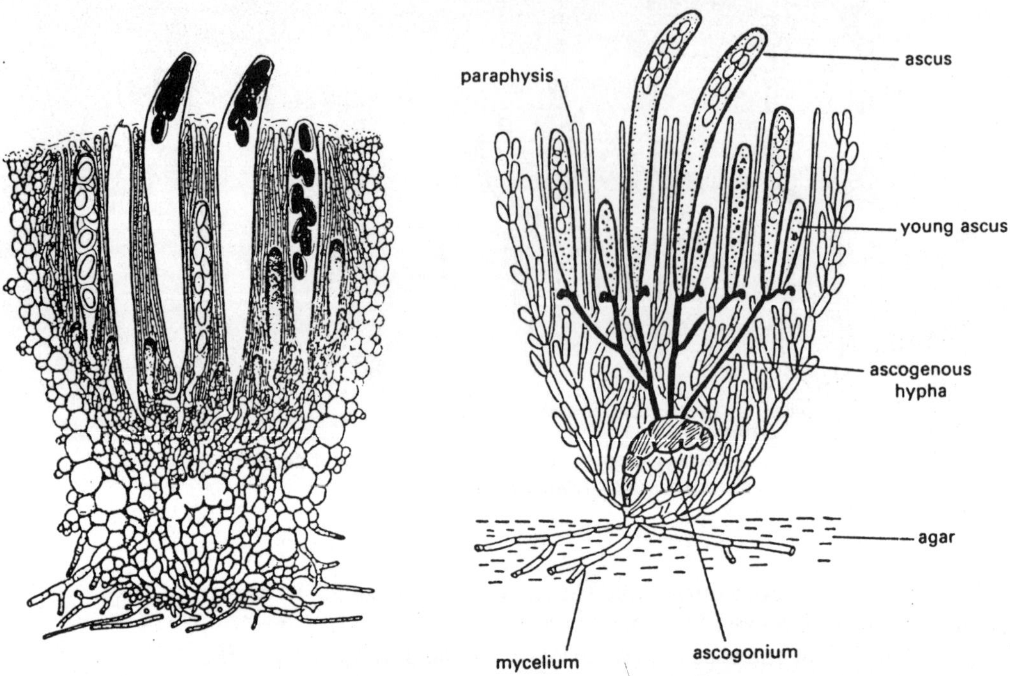

Fig. 4.5. *Ascobolus stercorarius.* Left, vertical section of an apothecium (after Corner) × 140. Right, diagrammatic interpretation; the female organ is shaded and the ascogenous hyphae are indicated by solid black lines.

In *Ascobolus stercorarius* a culture derived from a single ascospore fails to form apothecia. If a number of such cultures are produced, these can, as in *Mucor,* be sorted into 'plus' and 'minus' strains alike in appearance. In *Ascobolus,* however, both strains are hermaphrodite.

The female apparatus (Fig. 4.6) consists of a partially coiled hypha with a stalk region of few cells, a more swollen ascogonium and an apical trichogyne of 2–3 cells. All the cells are multinucleate. The male cells are borne in chains on erect hyphae. Although they are sexual cells, they can also behave as asexual spores and germinate to give a new mycelium. The sexual process can be observed when male cells of one mating type are placed on a culture of the opposite kind. On the surface of the agar the trichogyne grows towards and fuses with a nearby male cell, and the nuclear contents from this pass from cell to cell into the ascogonium. Here, it seems, there is no nuclear fusion, but a close association of two nuclei. The pair is rapidly replicated by conjugate division, the two nuclei dividing simultaneously to give two daughter pairs, and so on.

Following union of the female organ with a male cell, ascogenous hyphae develop from the ascogonium (Fig. 4.5). In these the nuclei are in pairs and their conjugate division continues. The ascogenous hyphae are thus dikaryotic, that is to say they contain nuclei of two genetic types.

Eventually from the tips of ascogenous hyphae asci are formed in a

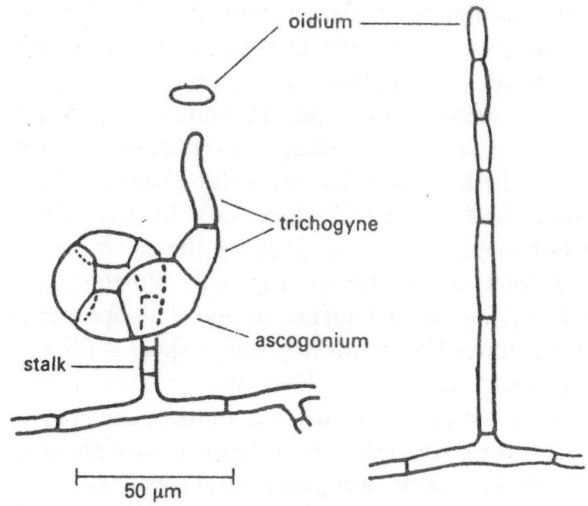

Fig. 4.6. *Ascobolus stercorarius.* Sexual apparatus. Left, the trichogyne of the female branch is growing towards an oidium (acting as a male cell) on the surface of the agar. Right, chain of aerial oidia produced from the prostrate mycelium on the agar.

characteristic manner (Fig. 4.1). The end of a hypha bends over to form a crook which is binucleate. Then conjugate division occurs to give four nuclei followed by wall formation delimiting: a downward-facing uninucleate, apical cell; a penultimate, binucleate ascus-initial cell; and a uninucleate one below that. Between this cell and the apical one open communication is established and nuclear migration restores the binucleate condition in the apical cell. This grows up to form a new crook which behaves like the original one. In the ascus-initial cell the two haploid nuclei fuse forming a diploid one which immediately undergoes meiosis. As a result four haploid nuclei are produced, and each of these divides again by mitosis giving eight. While these nuclear events have been occurring, the ascus has been elongating. Next, around each nucleus an ascospore is organized, but some protoplasm (epiplasm) is left outside the spores. Initially the epiplasm is rich in glycogen staining chestnut-red with iodine. The ascus continues to enlarge and a central vacuole develops in which the spores are suspended in an apical position. The epiplasm is finally reduced to a thin layer lining the stretched ascus wall and the glycogen reaction is lost. Presumably the glycogen, a polymer of glucose, is converted to the monomer. The result is that a considerable hydrostatic pressure is developed in the ascus by osmosis because of the high sugar content. Eventually the ascus bursts shooting its spores to a distance of 5–10 cm.

We have been discussing the development of the dikaryotic ascogenous hyphae from the ascogonium, but in the construction of the whole apothecium monokaryotic hyphae from the stalk region of the female organ also play an important part. The apothecium, is, in fact, a compound structure consisting of an interwoven mixture of the two kinds of hyphae. Most of the tissue of the apothecium is provided by the monokaryotic hyphae, and in the hymenium itself only the asci are derived from the ascogenous hyphae, the paraphyses being ultimate branches of the vegetative, monokaryotic system (Fig. 4.5).

Peziza and *Ascobolus* are two of the many genera belonging to the Pezizales in which the ascus dehisces by a lid. In another big order, Helotiales, dehiscence is by a pore. A common example is *Trichoglossum hirsutum* found on lawns on acid soil. The black apothecium is club-shaped with the hymenium covering the upper half of the club (Fig. 4.4). In most Discomycetes the ascospores are egg-shaped, but in *T. hirsutum* the ascus contains a sheaf of eight rod-like, parallel, septate spores (Fig. 4.2). Further, instead of being discharged together, they are shot out one by one. The ripe ascus dehisces by a minute apical pore. Immediately this happens, a spore is forced into the pore temporarily stoppering the ascus. Then this is shot away, due to the hydrostatic pressure within the ascus, but immediately the ascus is again briefly stoppered by the next spore, and so on until the eight spores have been discharged. The whole process occupies several seconds.

Although without an exposed hymenium, the truffles, placed in the order Tuberales, are included in the Discomycetes. They have underground ascocarps, but, nevertheless, seem to have been derived from normal discomycete ancestors. The ascospores are not actively discharged and appear to rely for dispersal on the ascocarps being located and eaten by rodents. Members of Tuberales are probably rather common, but are infrequently collected. To find them involves raking in the soil below the leaf-litter in woods, and being familiar with their appearance. Apart from *Tuber aestivum* and *T. melanosporum*, most members of the order are rather small and inconspicuous.

In *Genea* (Fig. 4.7) the ascocarp is clearly a modified apothecium. A vertical section shows a distinct hymenium of parallel, elongated asci, each with a row of eight ascospores. The asci are interspersed with paraphyses, but these grow beyond the palisade of asci and intertwine to form a pseudoparenchymatous tissue. The asci never burst.

More familiar are species of *Tuber*. *T. melanosporum* and *T. aestivum* are the edible truffles so highly prized by gourmets. *T. aestivum* (Fig. 4.8) is not uncommon in deciduous woods and its ascocarps may be found just below the leaf-litter. A truffle, which can weigh over 200 g, is a firm structure of an irregular oval form having a hard, black, warted outside. A section gives no suggestion of discomycete ancestry. It is only by considering seemingly intermediate types that this idea can be sustained. In the solid mass of the ascocarp, there are anastomosing sterile veins ramifying through the pseudoparenchymatous tissue which is evenly studded with spherical asci (Fig. 4.8). There is no organization of asci into a definite hymenium as in *Genea*. Further, the asci are not elongated, but are oval or spherical and each contains a reduced number of ascospores ranging from one to four. The spores are relatively large with beautifully ornamented walls.

In France truffles are so much prized that dogs and pigs are trained to locate them. They find the truffles by the strong and unpleasant smell which, however, is emitted only when the ascocarps are mature. If a ripe specimen of *T. aestivum* is brought into the house and left there overnight, the whole place

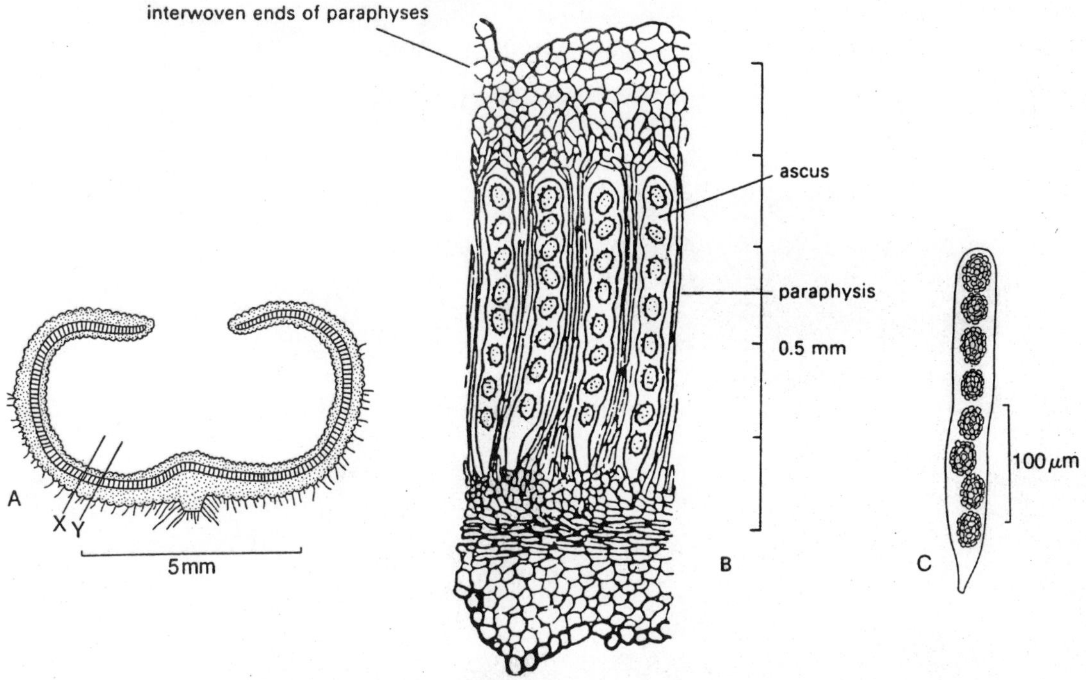

Fig. 4.7. *Genea hispidula*. A, vertical section of apothecium. B, details of a part of A between lines X and Y. C, ascus with mature ascospores.

stinks, and, even after it has been banished, it takes days to get rid of the smell. In nature rodents dig up the ripe truffles and eat them, the spores presumably passing uninjured through the alimentary canal. They are probably voided with the faeces in a viable condition. However, nothing definite is known about this, as mycologists have not yet succeeded in germinating the spores.

Ascomycotina is not the only division of fungi in which underground reproductive structures have developed. They also occur among Basidiomycotina (e.g. *Hymenogaster*, see p. 99 and in Zygomycotina e.g. some Endogonales, see p. 40).

PYRENOMYCETES

The Pyrenomycetes or flask-fungi are numerous and show great variation in size and structure. The asci, instead of being organized in the exposed hymenia of apothecia, are enclosed in a flask-shaped perithecium opening to the outside by a narrow neck canal through which only one ascus can extend at a time. Individual perithecia are always small. In some genera they occur singly, while in others they are grouped together in a common stroma, which may be quite large.

Sordaria may be considered as an example of a pyrenomycete with solitary perithecia. *S. fimicola*, like *Pilobolus* and *Ascobolus*, is a regular member of the

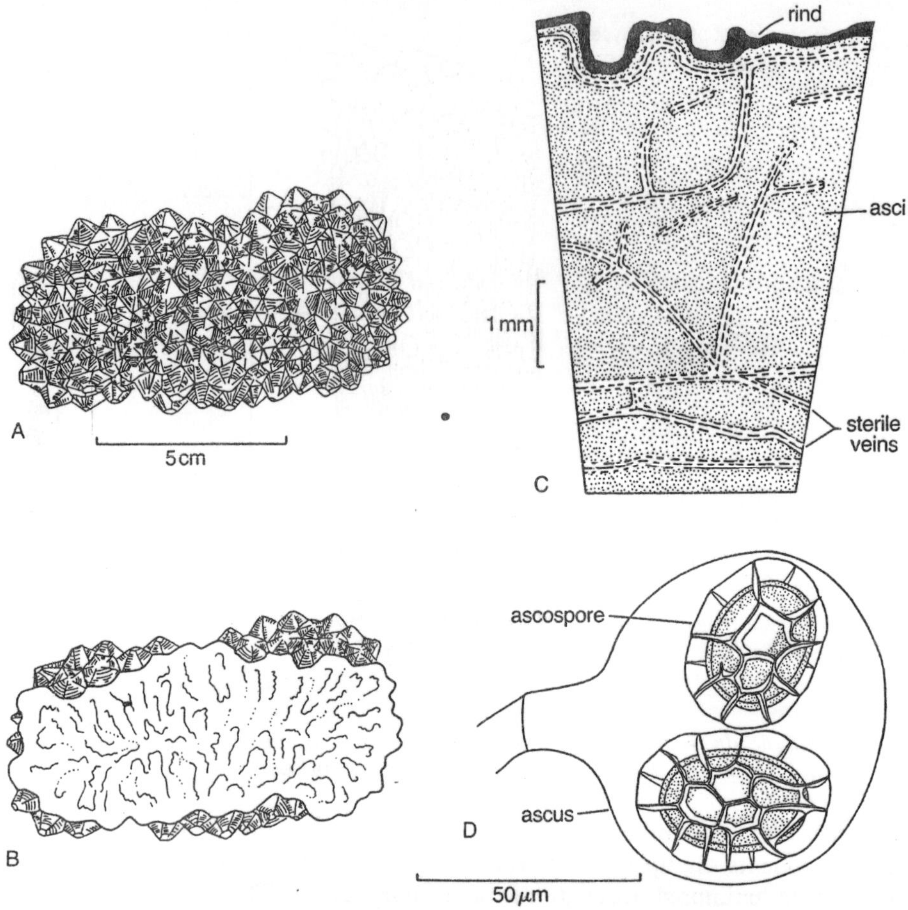

Fig. 4.8. *Tuber aestivum*. A, whole ascocarp. B, section showing the veins of sterile tissue. C, small part showing the rind, veins of sterile tissue and asci each indicated by a black dot. D, single ascus with two ascospores.

fungi that develop so consistently on the dung of herbivores. It grows readily on nutrient agar and has been extensively used in experimental studies. In agar cultures the perithecium, about 0.5 mm high, has its lower part immersed in the jelly with its neck projecting into the air (Fig. 4.9). The interior of the perithecium is occupied by a hymenium of asci at all stages of development. In spore discharge one ascus, which happens to be a little ahead of the rest, elongates up a narrow canal, lined by delicate hyphae, leading to the mouth or ostiole of the perithecium. Once the tip reaches the outside, the ascus bursts shooting the ascospores to a distance of up to 10 cm. The empty ascus retracts into the perithecium and quickly dissolves. Following the discharge of one ascus, another elongates, and so on in orderly succession, a few minutes usually elapsing between the discharge of successive asci (Fig. 4.10). The process of puffing, so common in Discomycetes, cannot occur. We have seen that in *Peziza* and *Ascobolus* the individual asci are positively phototropic so that the ascospores are shot towards the light. In *Sordaria* the ascospores are

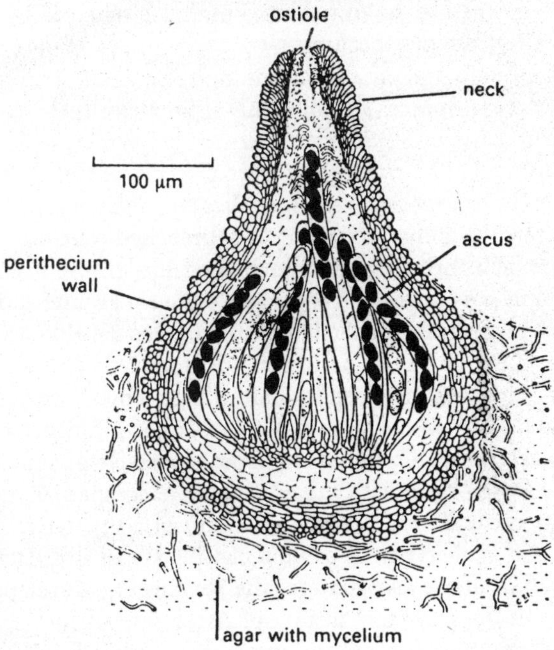

Fig. 4.9. *Sordaria fimicola*: vertical section of a perithecium.

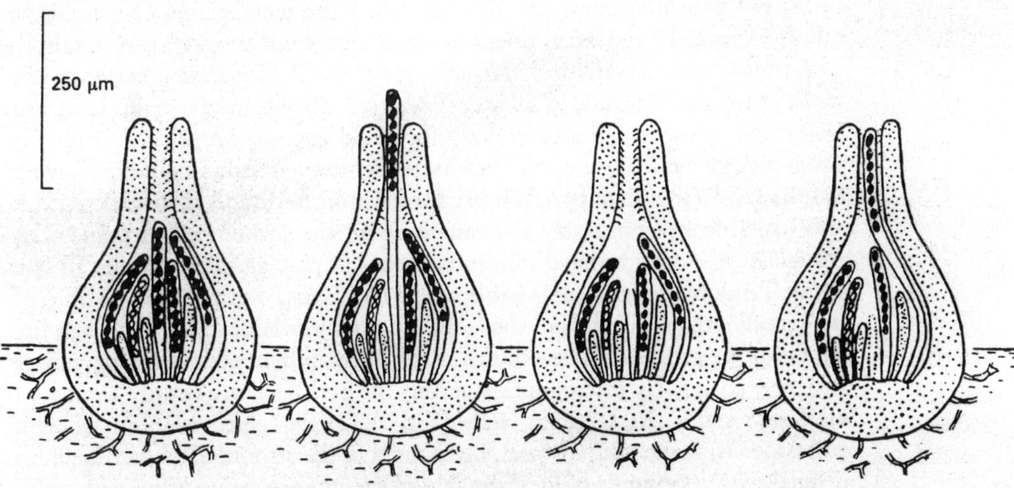

Fig. 4.10. Diagram of discharge in *Sordaria*. On the left the tip of an ascus is just entering the neck canal; a few minutes later the tip of the ascus protrudes beyond the ostiole; the next instant it bursts and the empty ascus retracts into the perithecium; some minutes later the empty ascus is breaking down and another ascus is elongating up the neck canal.

also discharged in that direction, but this is achieved by the phototropism of the neck of the perithecium.

S. fimicola is homothallic, a fruiting culture developing from a single spore. Further, ascospores are the only spores formed. However, a different situation exists in *Neurospora*, a genus very similar to *Sordaria*. Two species, *N. sitophila* and *N. crassa*, have been used extensively in genetical research. In addition to ascospores, *Neurospora* has two distinct types of conidia; relatively large macroconidia occurring as branched chains, and much smaller microconidia formed in unbranched chains. In the macroconidial condition the fungus is a pink mould often seen on burnt ground and charred vegetation. *N. sitophila* is heterothallic but female organs develop on both strains. Where compatible strains meet perithecia are produced. The female organ consists of a stalk, an ascogonial region and a long, septate trichogyne which is often branched. The whole, except for the projecting trichogyne, is encased in a sheath of hyphae. The female apparatus can develop into a perithecium only if the trichogyne fuses with a male element of compatible mating type. This can be a microconidium, a macroconidium, or a vegetative hypha. It should be noted that both types of conidia can behave as ordinary spores. Following fertilization of the ascogonium, the perithecium develops in much the same manner as does the apothecium of *Ascobolus*.

Although the ascospores in Pyrenomycetes are usually actively discharged, there are some members of the class in which this does not occur. Notable in this connection are species of the large genus *Chaetomium* which have a pronounced ability to utilize cellulosic substrates such as damp paper. In species of *Chaetomium* the perithecium is richly beset with stiff, hair-like hyphae. An example, *C. globosum*, is shown in Fig. 4.11. In all species of the genus the asci break down while still within the perithecium, producing a mass of spores mixed with mucilage which oozes out through the ostiole.

We may now consider a stromatal pyrenomycete. *Daldinia concentrica* is a striking example occurring as a common saprotroph on dead limbs of ash (*Fraxinus*). The mycelium colonizes the wood and the perithecial stroma is produced on the surface of the bark in September. It forms rapidly, growing to full size in a month or two. It is hard and hemispherical. At first dark grey, it ends coal-black. Commonly a stroma is about the size of half an apple. Cut vertically, it is seen to have concentric zones of pale and dark tissue. In late autumn tiny black dots, just visible with a hand lens, indicate the positions of the female organs. It is from these that the perithecia develop, although the details of the process are not known. Slow development occurs during winter and early spring. By the beginning of May the perithecia are mature and spore discharge starts. In sectional view (Fig. 4.12) the perithecia can be seen embedded in the extremely hard, black crust of the stroma. Each perithecium is lined with thousands of asci, the rest of the interior being filled with jelly. Communication with the outside is by a short neck canal. In general spore discharge is as in *Sordaria*, but the spores are shot to a distance of only 0.4–1.5 cm. While in *Sordaria* discharge is mainly by day, that in *Daldinia* is nocturnal. If a median vertical slice, about half a centimetre thick, is cut from an active stroma, laid horizontally on a sheet of glass and left overnight, a

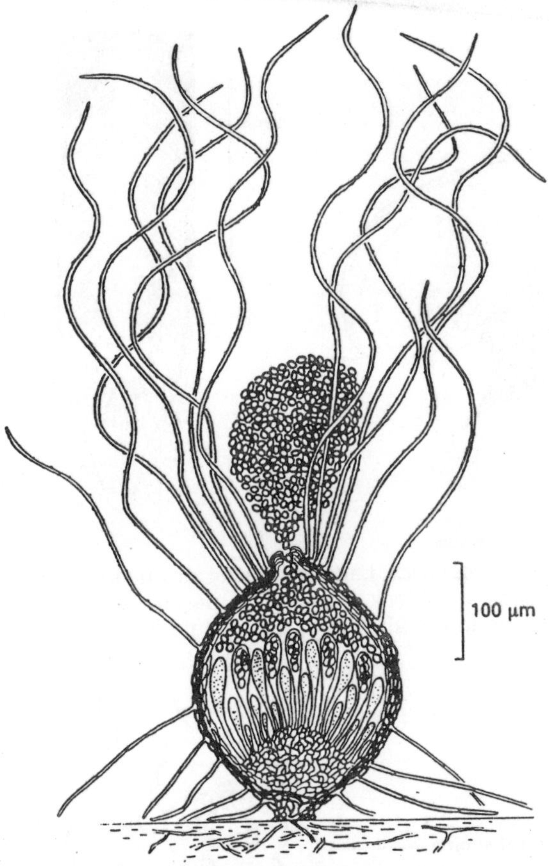

Fig. 4.11. *Chaetomium globosum*. Vertical section of a perithecium growing on agar. Asci break down within the perithecium and the spores ooze out through the ostiole.

dense deposit of black discharged spores is formed as a band about a centimetre wide parallel with the surface of the stroma, and separated from it by a spore-free zone a few millimetres in depth.

Spore discharge in Ascomycotina is dependent on an adequate supply of water since the bursting of turgid cells is involved. For this reason in the great majority of flask-fungi spore liberation can occur only under damp conditions, usually after wetting by rain. *Daldinia* is unusual in having a usable supply of water in its stromatal tissue. If a stroma is collected in May or June and placed in a dry room without any external supply of water, nightly discharge continues for weeks until the water reserve is at last exhausted. In *Daldinia* spore discharge normally continues from May to September. Then the exhausted stroma dies, an individual having but one season of activity.

Mention may also be made of another stromatal type, *Nectria cinnabarina* (Fig. 4.13). This is the coral-spot, a common saprotroph on twigs and branches of many kinds of tree and shrub. Two types of reproductive

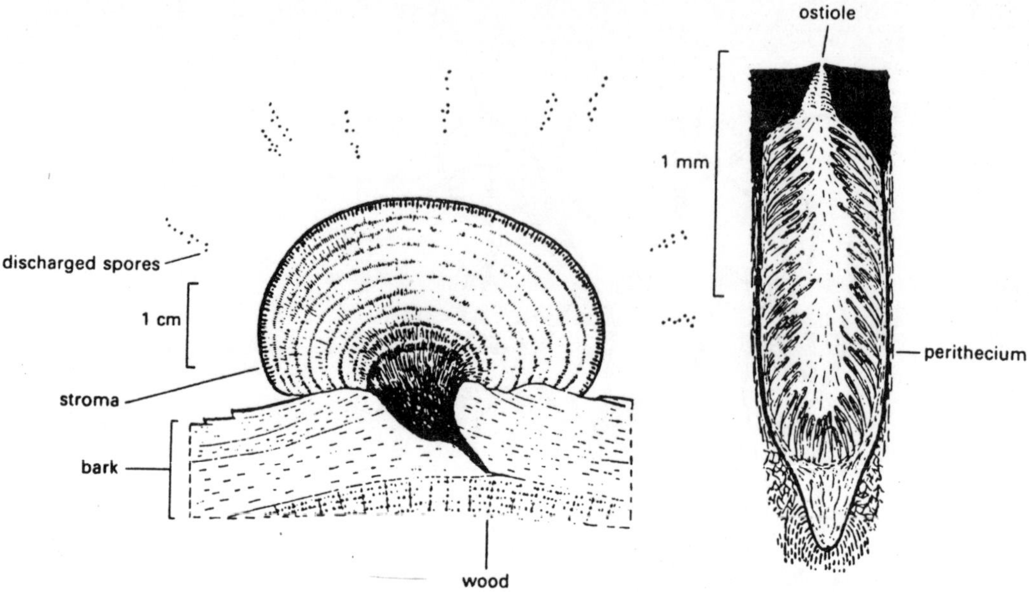

Fig. 4.12. *Daldinia concentrica*. Left, perithecial stroma in sectional view, the black dots just inside the black crust being perithecia; some spores just discharged are shown in the air. Right, a single perithecium.

Fig. 4.13. *Nectria cinnabarina*. Twig bearing stromata as seen under a lens. Each perithecial stroma is a group of dark red perithecia. Each smooth conidial stroma is coral pink. Some stromata are mixed. (After Tulasne)

structure are formed, both breaking through the bark and occurring in great numbers on the surface. The more conspicuous are the conidial stromata, coral-pink when dry, each a cushion about 2 mm across. Under wet conditions the surface of the stroma becomes a slimy mass of conidia which are dispersed by splashing rain. The perithecial stroma is about the same size, but is dark red and under the lens looks like a raspberry, each segment being a perithecium. In sectional view it can be seen that the perithecia are not deeply embedded in the stromatal tissue, but project from it. Often both types of reproductive structure are to be found on the same twig, and a few stromata are in part perithecial and in part conidial.

In its spore discharge, *Nectria* is a typical drought-enduring fungus. In dry weather the somewhat fleshy perithecia shrivel. After rain they quickly recover and spore discharge begins again. There is no capacity to continue discharge during dry periods.

PLECTOMYCETES

Erysiphales

Although Erysiphales and Eurotiales are classified together in Plectomycetes, all they have in common is the production of a closed ascocarp or cleistothecium. This is probably a structure that has been evolved again and again within Ascomycotina, and there are no good grounds for believing that the two orders are closely related. They will here be considered separately.

Erysiphales are obligate biotrophic parasites of the leaves and sappy stems of a wide variety of plants. They are referred to as 'powdery mildews' because the affected parts have a white, mealy appearance produced by masses of conidiophores. Rose mildew (*Sphaerotheca pannosa*), hop mildew (*S. humuli*), mildew of cereals (*Erysiphe graminis*), vine mildew (*Uncinula necator*) and apple mildew (*Podosphaera leucotricha*) are important examples.

Sphaerotheca may serve as an example. The conidium germinates on the surface of the leaf. In most fungi spores can germinate only in water, but in the powdery mildews not only can the conidia germinate on a surface free from water, but also at humidities that are remarkably low. Following germination, a branched, septate mycelium of uninucleate cells is produced on the epidermis of the host obtaining its nourishment through haustoria developed within the epidermal cells without destroying them. The haustorium has a nucleus and is limited by a very thin membrane closely applied to the cytoplasmic lining of the epidermal cell. This cytoplasm, though pushed out of shape by the haustorium, is not disrupted. The haustorium presents a considerable surface through which soluble substances can be absorbed from the living host cell and translocated to the developing mycelium and conidial chains on the leaf surface (Fig. 4.14).

In developing the mycelium on the surface of the host, powdery mildews are unlike other obligate parasites such as downy mildews (Peronosporaceae) and rusts (Uredinales) in which the mycelium is within the host tissue. From the

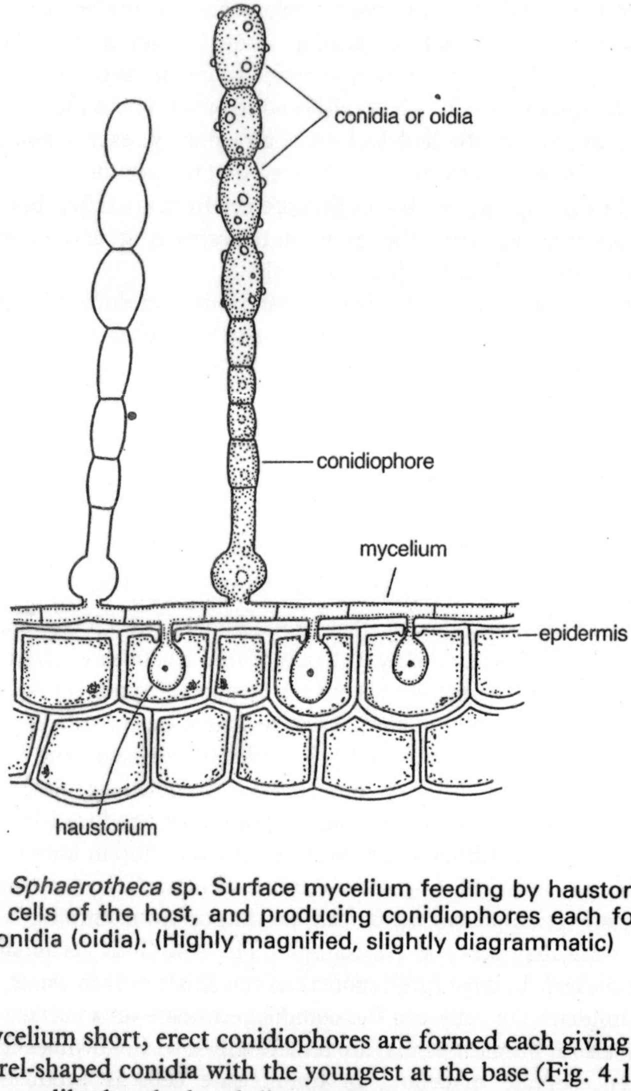

Fig. 4.14. *Sphaerotheca* sp. Surface mycelium feeding by haustoria in the epidermal cells of the host, and producing conidiophores each forming a chain of conidia (oidia). (Highly magnified, slightly diagrammatic)

surface mycelium short, erect conidiophores are formed each giving rise to a row of barrel-shaped conidia with the youngest at the base (Fig. 4.14). Ripe conidia are readily detached and dispersed by wind.

Spread of the disease by conidia goes on throughout the summer. This is the epidemic stage. Later cleistothecia are produced. Under a hand lens these appear as minute black dots scattered among the surface mycelium. The cleistothecium is spherical, around 0.1 mm in diameter, completely closed and when ripe is dark brown or black. Within is a single ascus. From the outer wall of the cleistothecium distinctive hyphae are formed known as appendages (Fig. 4.15).

The cleistothecium develops following a sexual process. It represents the resting, hibernating stage of the pathogen. During winter the single ascus within remains small. The ascospores are already delimited, but not completely developed, and there is considerable epiplasm around them. In

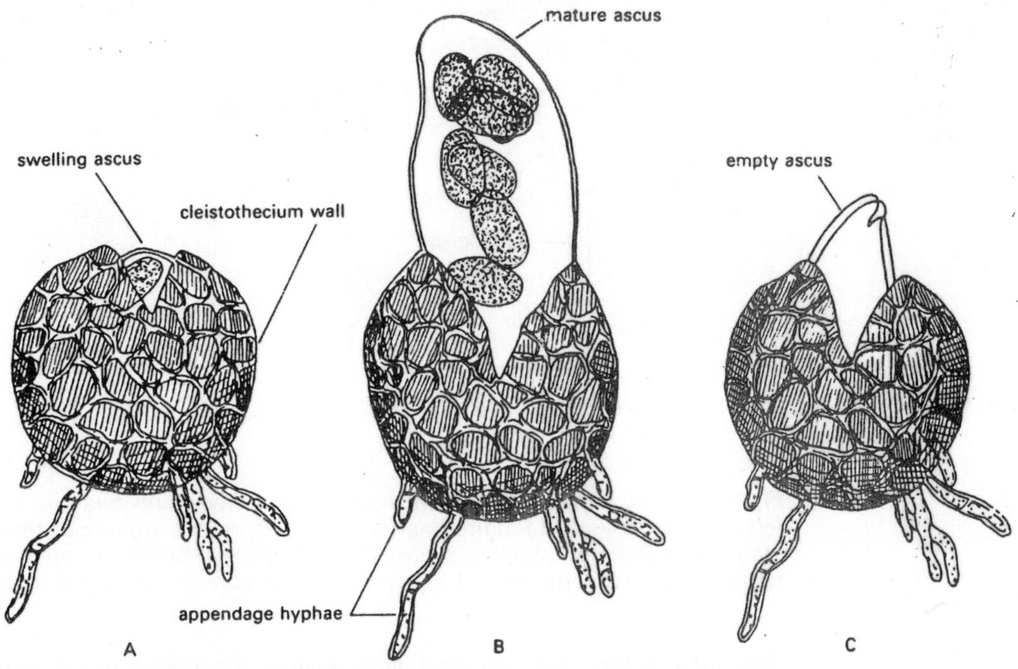

Fig. 4.15. *Sphaerotheca mors-uvae.* Discharge of ascospores from a cleistothecium. On the left the swelling ascus is just bursting the wall of the cleistothecium; in the middle the fully swollen ascus is about to discharge its spores, and on the right the ascus immediately after discharge (highly magnified; after Salmon)

late spring activity recommences. The spores become fully formed and the visible material of the epiplasm disappears, probably owing to conversion to sugar, resulting in a decrease in the osmotic potential. The ascus absorbs water, swells, rupturing the wall of the cleistothecium and extending beyond it. It then bursts, throwing its spores into the air. These may cause the first infections of the new season. However, although over-wintering is typically by cleistothecia, a number of powdery mildews (e.g. rose mildew) may pass the winter as viable conidia within the dormant buds of the host plant.

In powdery mildews the conidial stage is much the same in all the genera which are distinguished on the basis of the cleistothecium. The two features of taxonomic importance are the number of asci present (one in *Sphaerotheca* and *Podosphaera*, several in the other genera) and the form of the appendages.

Powdery mildews are mostly controlled by fungicidal dusts or sprays, especially those which are sulphur-based, but there are many systemic fungicides now available that are just as effective.

Eurotiales

Eurotiales is an interesting order of moulds in which the conidial stage is much more prominent than the ascus stage which in most species is rarely seen. To

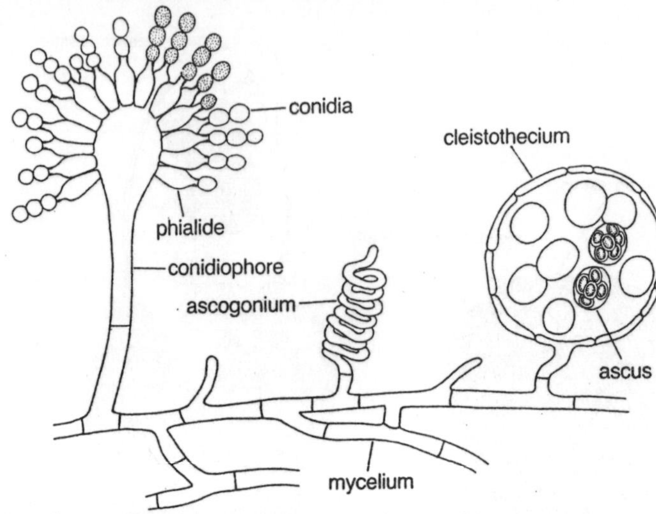

Fig. 4.16. *Aspergillus (Eurotium) herbariorum*. Mycelium bearing, on the left, conidial stage (in longitudinal section) and, on the right, cleistothecial stage (also in longitudinal section). The coiled aerial hypha between is the female organ (ascogonium). (Highly magnified, somewhat diagrammatic)

this order belong the two very large genera *Aspergillus* and *Penicillium* containing species of great economic importance.

There is a problem of nomenclature in relation to *Aspergillus*. It is a generic name reserved for the type of conidiophore illustrated in Fig. 4.16. It is an anamorph (see p. 55). Not all species of *Aspergillus* possess a teleomorph. If one is formed, it is usually of the type placed in the genus *Eurotium*.

Aspergillus (Eurotium) herbariorum (Fig. 4.16) is a well-known mould. The conidiophore is erect and unbranched, and its apex is swollen into a globular head. From this radiate a number of phialides. The phialide is a special type of structure found in a large number of conidial fungi. It tends to be wider in the middle and to taper slightly to both apex and base (Fig. 4.17). A conidium is budded from the apex and is not separated by a basal wall until it is fully grown. Then another develops from the phialide, so that a succession of conidia is formed all originating at the same level. The conidia are dry and easily detached by wind. A culture of *A. herbariorum* is bright yellow, owing to pigmentation of the spores, but other species may be white, blue or black.

If *A. herbariorum* is grown at 15–25°C on a relatively dilute medium, only conidia are produced. However, on a medium with the sugar concentration stepped up to 10%, combined with a temperature of 25–30°C, cleistothecia are also formed. These are spherical and 150–200 μm in diameter. The cleistothecium has a cellular wall, and contains a large number of minute, spherical, eight-spored asci. These are not explosive. By the time the cleistothecium is fully mature, the walls of the asci have broken down leaving a mass of ascospores within. Only with the breakdown of the wall of the cleistothecium are these set free.

Most of the common blue moulds belong to *Penicillium*, a huge genus of 250 species which has been the subject of intense taxonomic study because of the

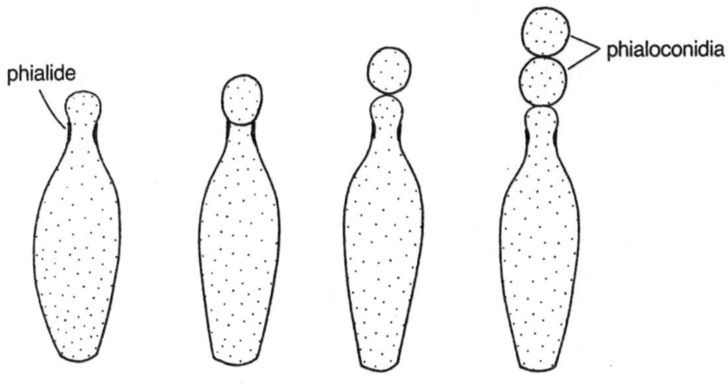

Fig. 4.17. Diagram of a phialide producing phialoconidia.

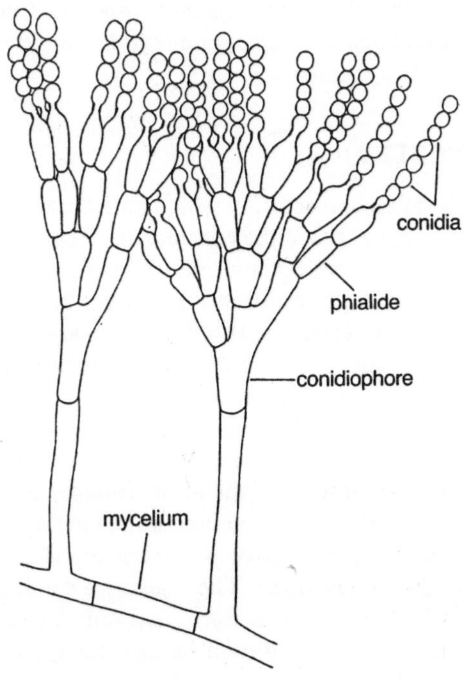

Fig. 4.18. *Penicillium* sp. Mycelium with conidiophores forming chains of conidia (phialoconidia). (Highly magnified)

economic importance of certain species. The conidial apparatus is characteristic (Fig. 4.18). It consists of a straight, septate stalk-hypha branching above to produce a group of parallel, terminal phialides. Each of these gives rise to a chain of conidia. The dominance of the anamorph is even greater than in *Aspergillus* and in most species the teleomorph is unknown.

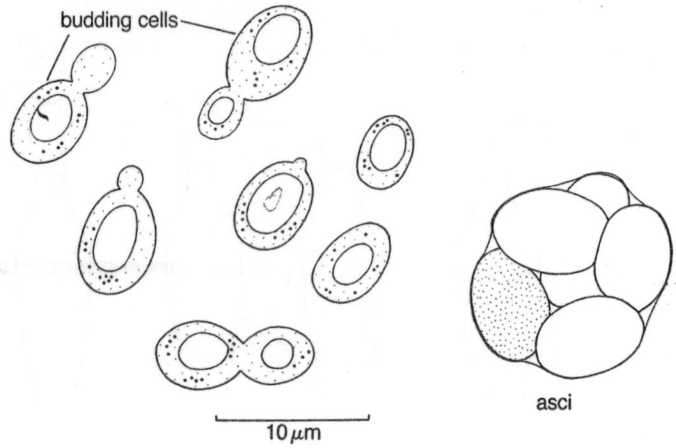

Fig. 4.19. *Saccharomyces cerevisiae*

Species of *Penicillium* and *Aspergillus* are extremely common saprotrophs in the soil and on damp decaying vegetable matter. Generally species of *Penicillium* are characteristic of temperate regions, while *Aspergillus* tends to predominate in tropical countries.

HEMIASCOMYCETES

Within the class Hemiascomycetes, in which no ascocarp is formed, only one family, Saccharomycetaceae, will be considered. Members of this family are the true yeasts, extremely common unicellular fungi found particularly on the sticky surface of ripe fruit, but also in many other habitats.

In considering yeasts in isolation, it is difficult to argue that they should be classified in Ascomycotina. However, there are mycelial genera which link yeasts with more typical Ascomycotina.

Another difficulty is that the yeast habit has been developed in other classes of fungi. Thus *Mucor racemosus* may form a yeast-like state (p. 33). Again, in Basidiomycotina the basidiospore of the smut *Ustilago*, transferred to nutrient agar, buds like a yeast and may be kept indefinitely in this condition. Further, the exceedingly common pink mirror-picture yeasts should almost certainly be classified in the Basidiomycotina. The vegetative condition represented by a budding yeast is merely a growth form, and only when the ascus stage is produced is it certain that a member of Saccharomycetaceae is involved.

The most noteworthy feature of yeasts is the production of ethanol from sugar under anaerobic conditions. *Saccharomyces cerevisiae* (Fig. 4.19) is the brewer's yeast. It is added to wort derived from barley in the production of beer. In making wine, however, yeast does not have to be supplied for it occurs naturally on the surface of the grapes. In a well-aerated sugary solution yeast rapidly multiplies by budding. The ovoid yeast cell has a conspicuous vacuole with a nucleus to one side of it. The cytoplasm may contain glycogen and fat as food reserves. In vegetative reproduction a bud develops at one end of the cell, nuclear division then occurs and a daughter nucleus moves into the bud which is finally pinched off as a separate cell. A daughter cell may sometimes bud before it has separated from its parent and so short chains may be formed, but basically yeast is unicellular.

Under suitable conditions most yeasts can be induced to form ascospores. As well as growing in liquid culture, they can readily be cultivated on nutrient agar forming characteristic slimy colonies. If actively budding cells from a colony on a rich medium are suddenly starved by being spread on a poor 'sporulating medium' on which budding is inhibited, ascus formation often occurs. Agar containing 0.5% sodium acetate has often been used as a sporulating medium. In *S. cerevisiae* the ordinary cell becomes the ascus. Its single diploid nucleus undergoes meiosis and around each of the resulting haploid nuclei an ascospore is organized. The ascospores are liberated only when the ascus wall autolyses.

LOCULOASCOMYCETES

Fungi in this class are distinguished by having bitunicate asci. These have two distinct, separable walls. The outer wall is thick and inextensible and the inner thin and extensible. At maturity, the outer wall ruptures at the apex allowing the inner wall to expand under hydrostatic pressure, generated by the conversion of glycogen to simple sugars lowering the osmotic potential of the ascus sap. The ascospores, which are usually coloured and multicellular, are released one at a time through an apical pore in the inner wall. This is illustrated by *Sporormia intermedia*, a common saprotroph on herbivore dung (Fig. 4.2), and by *Pleospora infectoria* (Plate 1c).

The ascocarp is usually a pseudothecium which closely resembles a perithecium in form and function but develops quite differently. There is a wide diversity of developmental patterns in the ascocarps of Loculoascomycetes. In pseudothecial formation, vegetative hyphae aggregate into a spherical structure with a heavily melanized outer layer. The whole resembles a sclerotium. Asci develop in a single cavity or loculus which forms within this. The pore or ostiole, through which the asci are discharged, arise by breakdown of pre-existing hyphae at a predetermined site. A neck of varying length then develops.

There are some 530 genera and 2000 species in the Loculoascomycetes. A number of genera contain well-known plant pathogens. *Venturia inaequalis* causes apple scab. It is the most important disease affecting apple leaves and fruit in the UK and is a good example of a quality reducing disease. The blemishing scab reduces the marketable value of the fruit. The fungus overwinters mainly as developing pseudothecia on fallen apple leaves. The cold of winter triggers ascus development and as the new apple leaves unfold in the spring, asci mature and discharge their ascospores up into the turbulent air after a good shower of rain. They impact on leaves and germinate. Bundles or strands of hyphae develop in a plate-like manner under the cuticle. At a latter stage the cuticle is ruptured by the growth of short, brown conidiophores, giving the scab a velvety, brown to black appearance. The conidia are dispersed by rain splash and wind from spring to early autumn spreading the disease. At leaf fall in the autumn the fungus grows out from the lesions and forms pseudothecial initials.

The conidia of two Loculoascomycetes form a very common part of the air-spora, especially in the late summer and autumn. *Alternaria alternata* (Fig. 6.2) is the anamorphic state of *Pleospora infectoria* which forms pseudothecia on dead overwintered grass and cereal stems. The conidia colonize and sporulate on the senescent leaves of many plants, but especially grasses. Conidia of *Cladosporium herbarum* (Fig. 6.2) dominate the dry air-spora of temperate regions in late summer. The air may contain as many as 12 000 conidia/m^3 and form 60–80% of the total air-spora. It is a phylloplane inhabitant and common primary saprotroph on a very wide range of plant materials, but especially leaves of herbs, grasses and deciduous trees. Its teleomorphic state, *Mycosphaerella tassiana*, is much less common but can be found on overwintered dead tree leaves in the late spring.

Both these conidial states are particularly prevalent on any senescent plant material which has been exposed to a damp period. For example, they, and others, can be seen as sooty flecks on the glumes of cereals in a wet harvest. They form part of a group of fungi often designated 'field fungi'. *Alternaria alternata* is also associated with the deterioration of cereal grains. It is also allergenic. Other species of *Alternaria*, such as *A. brassicae*, infect seeds of brassicas and are a major cause of losses in seed production.

Basidiomycotina

5

To basidiomycotina, with some 16 000 species, belong most of the large and conspicuous fungi found in fields and woods such as toadstools, bracket polypores and puff-balls. However, the division also contains many micro-fungi, especially two large and economically important orders of obligate biotrophic plant parasites: rusts and smuts.

The mycelium consists of branched, septate hyphae. Anastomoses frequently occur so that it becomes a three-dimensional network. Each septum has a central pore of a special kind known as a 'dolipore'. There is a strong tendency for the cells to be binucleate and 'clamp-connections' often occur associated with the cross-walls. Dolipores and clamp-connections, features characteristic of Basidiomycotina, will be considered in detail later.

The distinctive feature of the division is the basidium, a special type of sporangium which produces its spores, usually four in number, externally. Up to a point there is a close parallel between the development of the ascus and that of the basidium. The young basidium, like the ascus, has two haploid nuclei which fuse and the diploid nucleus thus formed undergoes meiosis. However, the resulting haploid nuclei then pass into the developing basidiospores which, unlike ascospores, are produced externally (Fig. 5.1).

Most Basidiomycotina, especially the larger ones, reproduce mainly by basidiospores, but quite a number, as in Ascomycotina, form accessory asexual conidia.

In the larger Basidiomycotina, as in some of the bigger Ascomycotina, the spore-producing units are aligned close together side by side to produce extensive layers or hymenia. Further, in most species, the basidium is, like most asci, a spore gun but with a very short range (0.1–0.2 mm).

A simplified classification follows, but this is not entirely comprehensive.

1 *Hymenomycetes*. The basidia are in extensive hymenia which are exposed when mature. Spores are shot from the basidium when ripe. This group includes agarics, boletes, bracket polypores and coral fungi. It is the largest class.
2 *Gasteromycetes*. The basidia are again in hymenia, but these are not exposed. The basidium is not a spore-gun, the basidiospores being set free

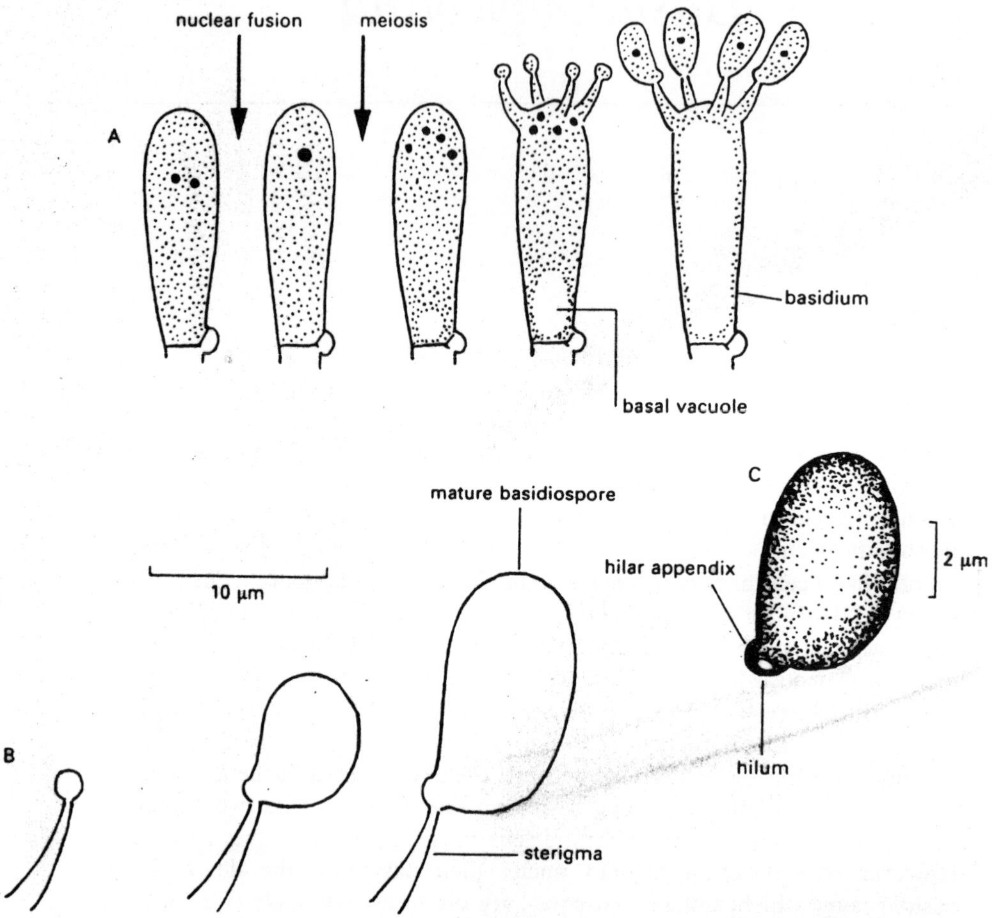

Fig. 5.1. A, diagram of stages in the development of a basidium. B, growth of a basidiospore at the tip of its sterigma in *Oudemansiella radicata*. C, *Agaricus* sp., liberated basidiospore as seen with the electron-microscope: the hilum appears as a white ellipse on the hilar appendix.

from the basidium by its breakdown. They line closed cavities within an initially closed basidiocarp. The basidiocarps remain closed until they disintegrate or they only expose their spores after they have matured and the basidia disintegrated. The group includes puff-balls, earth-stars, stink-horns and bird's-nest fungi.

3 *Teliomycetes*. This class contains two large and important orders of plant parasites. Unlike in the other two classes, basidiocarps are not formed. *Uredinales* (rusts). This is a large and important order of obligate biotrophic plant parasites. The basidium is transversely septate into four cells each of which forms a single basidiospore which is shot off. In these

fungi a number of different kinds of asexual spore may develop and the life-cycle is often complex.

Ustilaginales (smuts). This is also a group of plant parasites. The parts of the diseased plants where the smut reproduces break out into sooty black masses of teliospores. These on germination give basidia.

The first two groups are often referred to as 'higher Basidiomycetes'. In these the basidiocarp is fairly large and the basidium is unicellular. The other two groups are known as 'lower Basidiomycetes', most being without basidiocarps in the normal sense and with a septate basidium. Also included in lower Basidiomycotina are three small orders of gelatinous fungi, namely Tremellales, Auriculariales and Dacrymycetales. These jelly-fungi will not, however, be considered in this book.

HYMENOMYCETES

Hymenomycetes are divided into two orders: Agaricales and Aphyllophorales (Fig. 5.2). In the former the basidiocarp is fleshy and the hymenium covers gills (lamellae) or lines vertical tubes. In Aphyllophorales the basidiocarp is leathery or even woody and the hymenium is spread over a downward-facing and more or less smooth surface (*Stereum*), or covers an erect club or a system of vertical branches (*Clavaria* and other coral fungi), or less frequently the hymenium is on vertical teeth (*Hydnum*), or more usually (as in *Boletus* in Agaricales) lines vertical, downward-facing tubes (e.g. *Piptoporus* and other bracket polypores). The polypore habit has clearly been developed on more than one occasion in the course of evolution. The situation in *Boletus* seems to have been derived from fleshy gill-bearing ancestors, while the polypore habit in Aphyllophorales has apparently evolved independently.

'Toadstool' and 'mushroom' are old English words. Both are used by naturalists for the umbrella-shaped type of fungus with a discoid cap supported on a central stalk. A common popular conception is that toadstools are poisonous, while mushrooms are good to eat. However, the two words are synonymous in many contexts. The truth is that neither word has any accepted scientific meaning. In Britain 'mushroom' is mainly used by mycologists for a species of *Agaricus*, but in America the word tends to be applied to any gill-bearing fungus and to boletes.

In describing agarics (gill-bearing toadstools) the cultivated mushroom (*Agaricus bisporus*) is a useful species to consider because of its availability at all seasons (Fig. 5.3). However, this species is unusual in the organization of its feeding mycelium and in its breeding system. For these aspects of the biology of an agaric, it is more informative to consider *Coprinus cinereus* largely because it has been the subject of such intensive study.

The basidiocarp in *Agaricus* is a solid structure which, however, is composed entirely of interwoven hyphae. It is ephemeral, lasting only a few days, whereas the feeding mycelium, living on organic material in the soil, may endure for years. Recently it has been estimated, by DNA fingerprinting techniques, that the mycelium of *Armillaria bulbosa* may continue to grow for at least 1500 years by which time it may cover over 15 hectares.

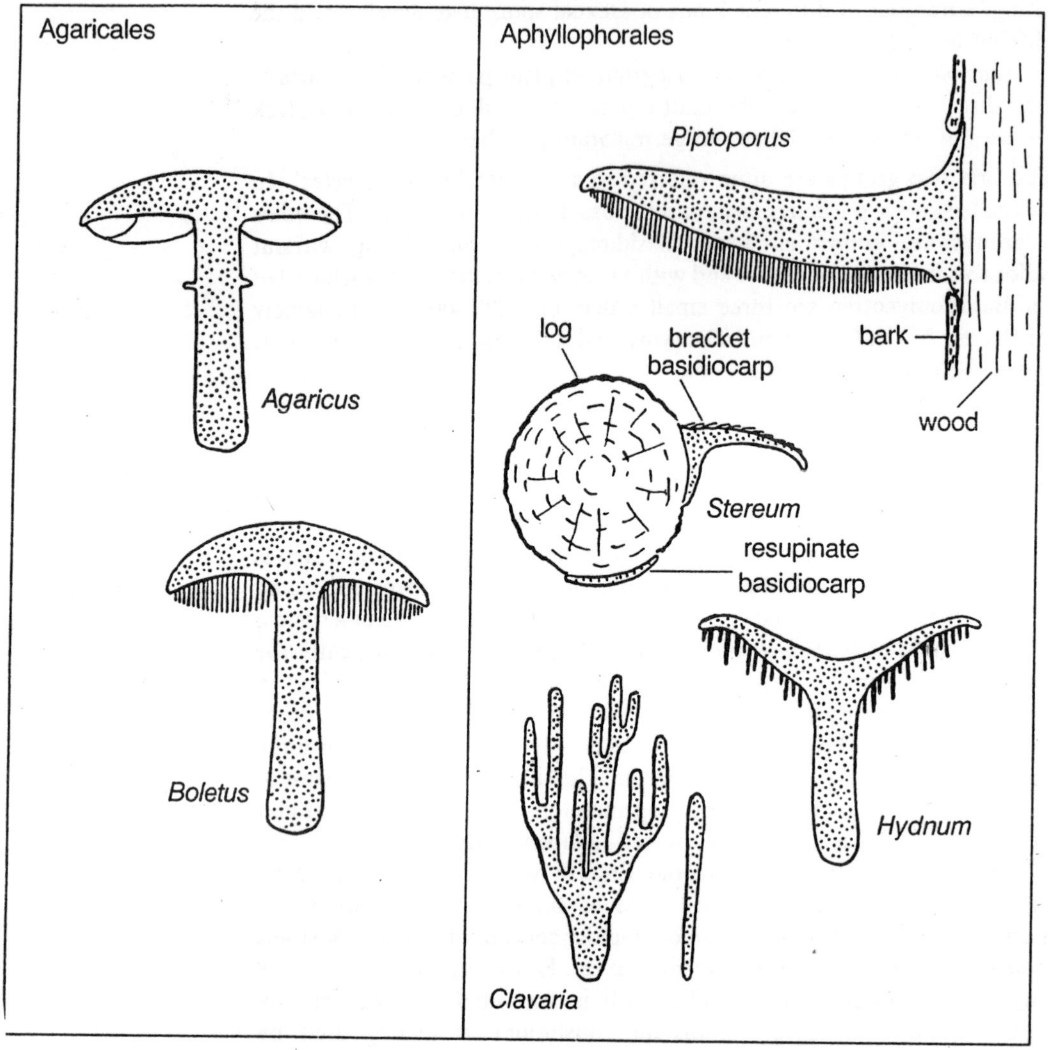

Fig. 5.2. Diagram of types of basidiocarps in Hymenomycetes.

The basidiocarp consists of a stout, rigid stalk (stipe) bearing a circular cap (pileus) from which the gills (lamellae) hang vertically (Fig. 5.3). If the stipe is removed and the pileus inverted, it may be seen that the gills are not all of the same length, they converge from the circumference towards the stipe. The longest gills stretch the whole way, but between two of these a medium-length one is inserted, and between each full-length gill and a medium-length one there is a still shorter one. This is illustrated (Fig. 5.4) for the common toadstool *Laccaria laccata* in which the gills are sufficiently far apart for this arrangement to be easily seen.

In a vertical, tangential section of the pileus, the hymenium is seen covering the surface of each gill. This spore-producing layer consists of elongated

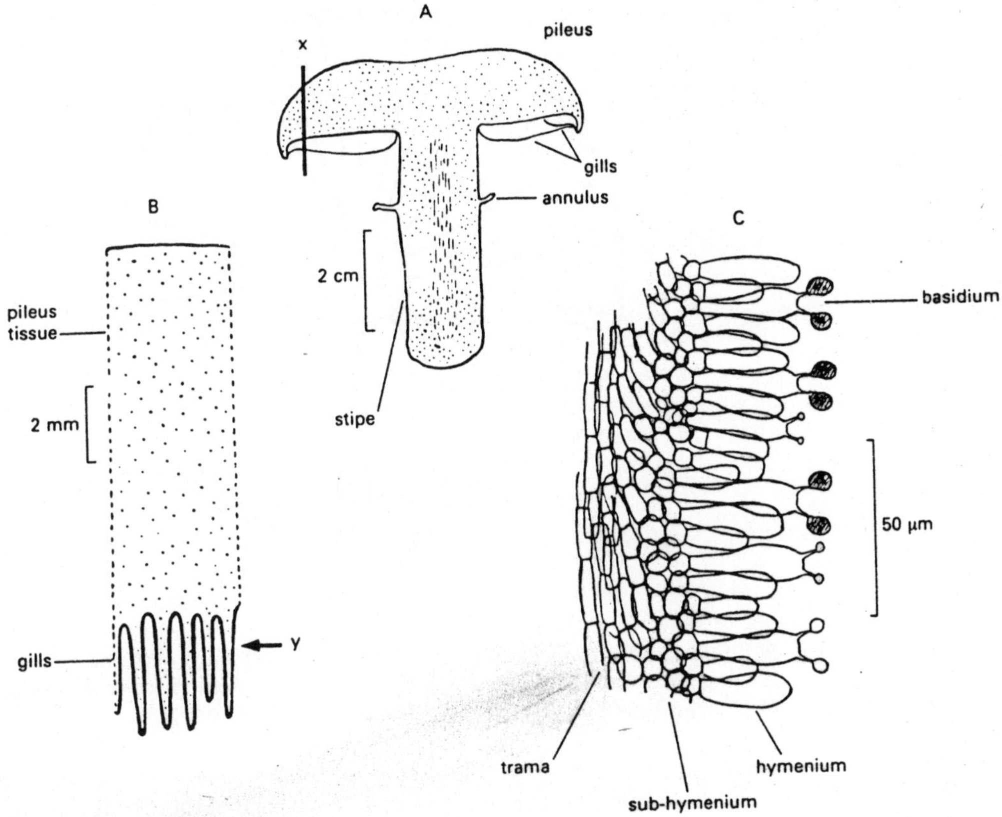

Fig. 5.3. *Agaricus bisporus*. A, vertical section of basidiocarp. B, tangential section of pileus along line X in A. C, small part of hymenium at Y in B.

basidia, at all stages of development, set at right-angles to the surface. It has often been stated that two types of element are present in the hymenium: basidia and sterile cells or paraphyses, but it seems that what were formerly taken for paraphyses are merely young basidia. In a few agarics (e.g. *Coprinus*), however, well-defined paraphyses occur. All parts of the gill in *Agaricus* are at roughly the same stage of development, and, if any small area is examined under the microscope, basidia at all stages are to be seen. Below the hymenium in the gill is a compact region (sub-hymenium) of pseudoparenchymatous tissue. The middle zone of the gill is occupied by parallel, elongated hyphae which constitute the trama (Fig. 5.3C).

Basidium development in Hymenomycetes is illustrated in Fig 5.1A. Fusion of the two haploid nuclei in the young basidium produces a diploid nucleus which at once undergoes meiosis to give four haploid nuclei. At this stage the basidium is full of granular protoplasm. Four little outgrowths (sterigmata) then arise at the top of the young basidium and the tip of each begins to inflate to form a basidiospore. Meanwhile a sap-filled vacuole develops near the base of the basidium and gradually enlarges as the spores

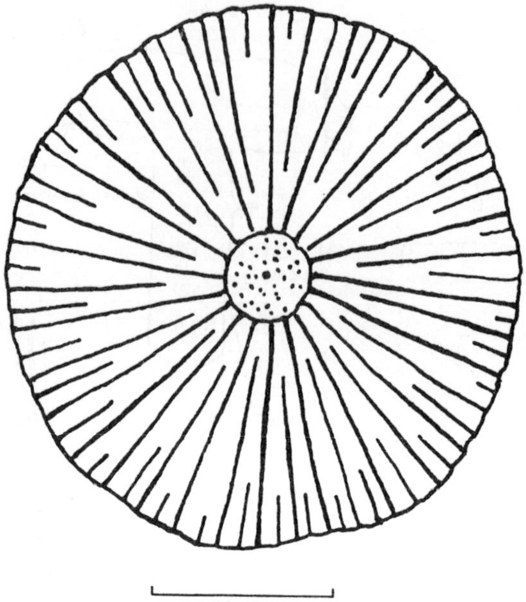

Fig. 5.4. *Laccaria laccata*. Underside of a basidiocarp with the stipe cut off leaving a central disc. There are 19 long gills stretching from circumference to stipe; 19 gills of three-quarter length; and 38 short gills.

swell. The enlarging vacuole seems to exert a piston action, driving the protoplasm into the developing spores. When each of these is nearly full-size, a nucleus squeezes into it through the exceedingly narrow tube of the sterigma. Finally the basidium appears to lose all its granular protoplasm, but, no doubt, an extremely thin layer of protoplasm continues to line its wall.

In some Hymenomycetes meiosis in the basidium is followed by a mitosis in the basidiospore so that they are binucleate.

In the cultivated mushroom (*Agaricus bisporus*) the basidium bears only two basidiospores (Fig. 5.3), each receiving two nuclei from the basidium. In the field mushroom (*A. campestris*), however, the basidia are four-spored.

The mature basidiospore is attached in an asymmetrical manner on its sterigma, and close to the point of attachment is a small projection known as the hilar appendix. It is this projecting hilar appendix which makes the liberated basidiospore such a characteristic object under the microscope. The actual attachment scar (hilum), being only 0.2 μm across, is too small to be seen under the light microscope, although it is easily seen with the electron-microscope (Fig. 5.1C).

In Hymenomycetes, a basidiospore starts as a minute spherical swelling of the tip of the sterigma. This then enlarges, but only on the side away from the longitudinal axis of the basidium, leaving the adaxial half of the swelling unexpanded as the hilar appendix (Fig. 5.1B). When the spore is fully grown there is, at first, cytoplasmic continuity with the basidium through the sterigma. However, this is ultimately interrupted, a situation established not

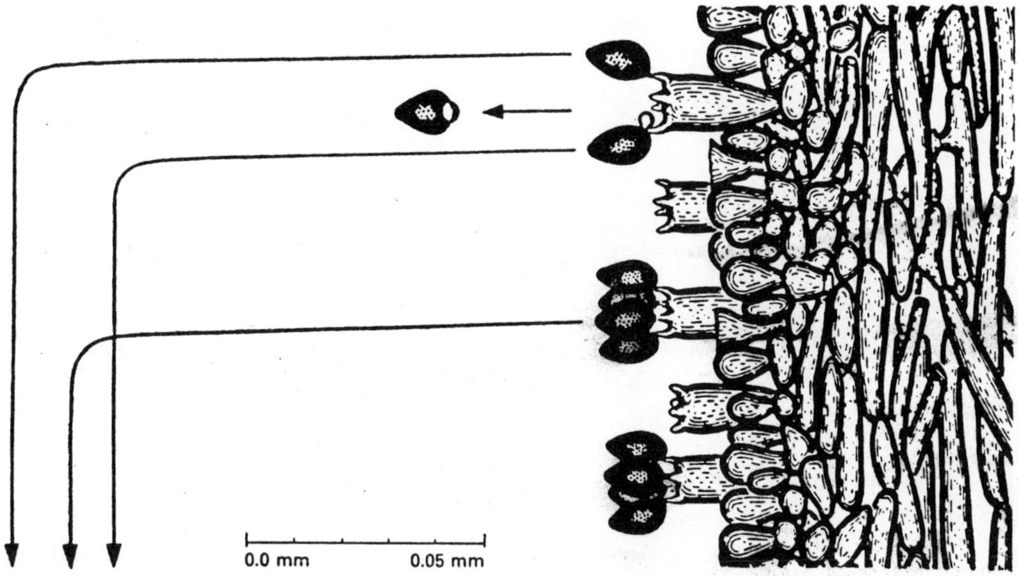

Fig. 5.5. *Panaeolus campanulatus*. Part of the hymenium in vertical section of the gill showing spore discharge. Trajectories of spores indicated. (After Buller)

only by transmission electron microscopy, but also by the fact that at this stage, but not earlier, the sticky spore can be removed by the lightest touch of a micro-dissection needle. Finally, the spore springs violently from its sterigma to a distance of 0.1–0.2 mm (Fig. 5.5). The four spores of the basidium are usually liberated in succession at intervals of a few seconds or even minutes.

The main features of the curious events associated with the active liberation of basidiospores were illustrated many years ago by Buller in his *Researches on Fungi*, but recently further observations have added significantly to the picture. The remarkable process is also to be seen in the liberation of aerial spores from 'mirror-image yeasts' and related filamentous fungi which all appear to belong to the Basidiomycotina. One of these fungi is *Itersonilia perplexans*, the study of which has added to an understanding of this kind of liberation. It has the advantage of having large spores produced singly under conditions that can easily be controlled. If watched continuously under the high power of the light microscope, a drop of liquid, the Buller drop, suddenly appears at the hilar appendix of the spore and, in the course of 20–40 s, grows to a size comparable to that of the spore itself. At the same time a blob of fluid makes its appearance on the surface of the spore (Fig. 5.6). Then, in a flash, the spore disappears travelling to a distance of about 1.0 mm. During these events the sterigma undergoes no visible change and, when it is vacated, its tip is without any adhering fluid.

Buller thought that the drop originated at or near the base of the spore, but there is now good evidence that it is formed by condensation from an

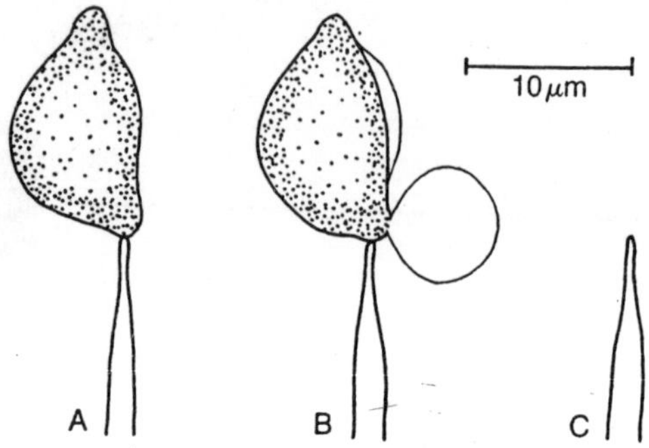

Fig. 5.6. *Itersonilia perplexans.* Single sterigma showing drop and blob. An interval of 40 seconds occurs between A and C.

atmosphere saturated with water vapour. Further, he did not observe the blob on the surface of the spore which has now been seen in a wide range of Basidiomycotina including toadstools, jelly fungi and rusts. It may well prove to be of universal occurrence in this kind of liberation but it is not so easy to observe as is the Buller drop.

Spores which are set free as described above are known as 'ballistospores', but it should be noted that not all basidiospores are ballistospores and not all ballistospores are basidiospores.

How ballistospores take off remains something of a mystery, but various mechanisms have been suggested including discharge by a liquid jet through the sterigma, sudden rounding-off of a cross-wall separating sterigma and spore, electrostatic repulsion and the sudden return flick of a sterigma momentarily bent inwards immediately before discharge. This last mechanism suggests that the slender, curved and tapering sterigma acts as a 'springboard'. At maturity the drop emerges from the adaxial side of the hilar appendix. As the drop enlarges, it adheres to the surface of the spore and spreads over the concave adaxial surface (Fig. 5.7). The adhesion of the drop drags the spore towards the axis of the basidium, putting the sterigma under tension and causing it to bend inwards. At this time a thin septum develops across the neck of the stalk and separation of the spore from its stalk occurs at this point. The spore breaks away releasing the tension on the sterigma which abruptly returns to its original position in a springboard manner. This projects the spore away from the basidium.

However, there is no concrete evidence in support of any of these mechanisms. A current theory which appears to be more consistent with the facts involves the spore springing away from its sterigma rather as one jumps from the ground by sudden upward transfer of the centre of mass of the body using muscular energy. It is suggested that when the Buller drop reaches a

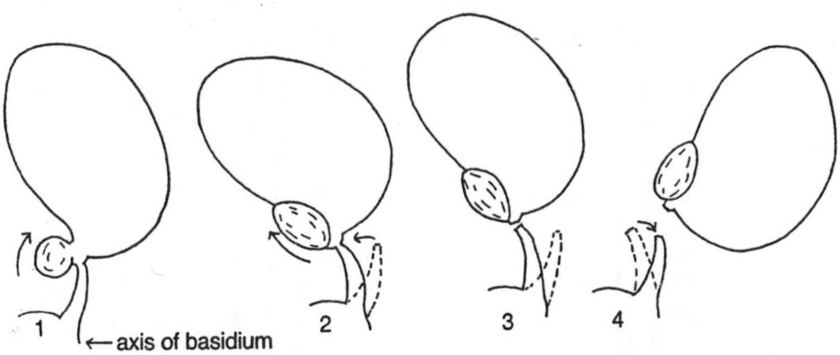

Fig. 5.7. Stages in the discharge of a single basidiospore by the springboard idea.

critical size, it instantly flows onto the side of the spore thus moving the centre of mass of the spore-plus-drop forwards, with the result that this takes off from a firm platform represented by the tip of the turgid sterigma. The energy involved is the surface energy of the drop and it has been calculated that this is adequate.

Mycologists have always referred to the 'discharge' of spores from the basidium, but if the mechanism just described is valid, the term would be inappropriate, since essentially they jump free.

A spore discharged by the drop-exudation mechanism is referred to as a 'ballistospore'. It has been noted that not all basidiospores are ballistospores, for those of Gasteromycetes are not discharged. Again, not all ballistospores are basidiospores for, as we shall see (p. 149), the aerial spores of the abundant mirror-image yeasts are discharged in the same manner as are the basidiospores of Hymenomycetes.

The gills of a mushroom are closely packed. Clearly the distance between opposing hymenial surfaces must exceed that to which the spores are shot, for otherwise the sticky basidiospores shot horizontally from the basidia would crash on to the opposite gill and stick fast. Actually the distance apart of the gills significantly exceeds the range of the basidial gun.

If the discharged spores are to escape from the basidiocarp, they must be able to drop freely into the air below the pileus. The whole organization of a mushroom appears to be related to the efficient liberation of the spores. The stipe has two functions: first, being quite strong, it holds the cap firmly so that it does not sway in the wind; and secondly it raises the pileus some way above the ground so that the spores have a reasonable chance of dropping into turbulent air in which they may be effectively dispersed. The stipe of an agaric is negatively gravitropic. This gives a rough vertical orientation to the gills. It is a coarse adjustment. But there is also a fine adjustment. Each individual gill is itself positively gravitropic, and, if it departs slightly from the vertical, growth movements occur in the gill bringing it again into the vertical plane. It is obvious that unless the gills are truly vertical escape of spores is impeded (Fig. 5.8).

From a fair-sized mushroom spores rain down at the rate of hundreds of

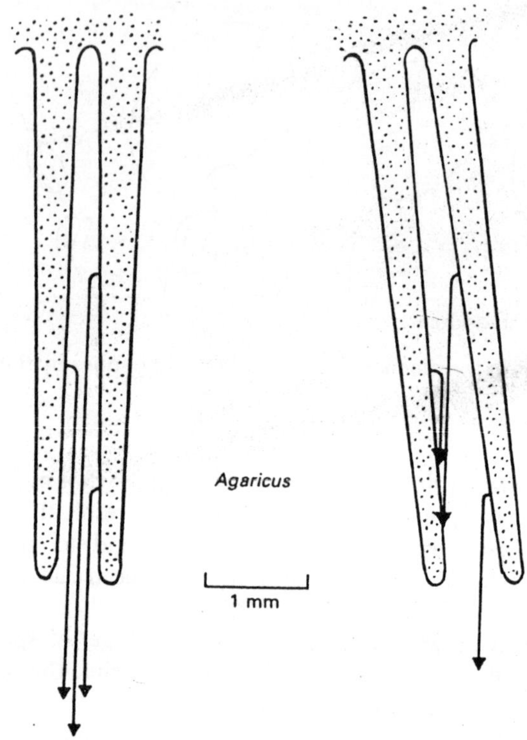

Fig. 5.8. *Agaricus campestris.* Part of a tangential vertical section of the pileus. Left: gills vertical. Right: gills 13° from vertical. Trajectories of some discharged spores shown.

thousands a minute during the entire two or three days of its active life. It is easy to demonstrate this rain of spores. To do this the stipe is removed from any agaric and the pileus is pinned in its natural position to a sheet of polystyrene foam loosely covering the mouth of a glass museum-jar with parallel sides. This is then placed in a darkroom and a strong parallel beam of light is passed horizontally through the jar at a level a few centimetres below the pileus. Through the beam the spores are seen falling at a steady rate. Although much too minute to be seen in their true form with the unaided eye, the individual spores scatter light and are visible as tiny falling stars.

If a pileus detached from its stipe is placed gills downwards on a horizontal sheet of glass, so that the gills are quite vertical, and left for a few hours away from any draughts, a spore-print is produced consisting of a series of radiating lines, picked out by millions of discharged spores, each representing the space between two gills. In the identification of agarics the spore-print is of some value. It may be white, cream, pink, purple, tan, brown or black. Usually in any one genus the colour of the spore-print is constant.

Under suitable conditions the dispersed spores may give rise to new mycelia, but wastage is enormous and the chances against a particular spore succeeding in its reproductive function are astronomical.

In agarics generally any tiny region of the surface of the gill has basidia at all stages of development. However, in the genus *Coprinus* (ink-caps) all basidia in such an area are at the same stage. *C. comatus* (Fig. 5.9) is one of the largest species. It is abundant in autumn on grassy areas rich in organic matter such as neglected lawns, road verges and abandoned rubbish dumps. A basidiocarp which is just starting to liberate spores has a characteristic appearance when split lengthwise. Each gill is white near the pileus tissue grading through pink and brown to black at the lower free edge. This is related to a gradient of hymenial development, the increasing pigmentation of the spores as they ripen determining the colour. In the gill the basidia are evenly spaced by sterile packing cells (paraphyses). If a gill is placed on a slide without added water and observed under the microscope, the beautiful pattern of four-spored basidia set in a pavement of paraphyses is easily seen (Fig. 5.9B). As soon as the ripe spores near the margin of the gill are discharged, the exhausted tissue of the gill undergoes autolysis. As this autodigestion proceeds, the gill gradually disappears and, in damp weather, fluid produced by autolysis flows to the margin of the cap from which it may drip as inky drops. When autolysis reaches the pileus tissue, this itself undergoes dissolution.

In the much smaller *C. cinereus* the organization of the hymenium is similar, but in addition the developing gills are kept apart by occasional very large cells of the hymenium known as cystidia (Fig. 5.9C).

C. cinereus is a coprophilous fungus. If freshly deposited balls of horse-dung are kept in the light under a transparent cover with adequate aeration, basidiocarps of this species usually develop in about ten days' time. They are variable in size. In some specimens the cap is only a few millimetres across; in others it may be several centimetres when expanded.

C. cinereus has been much studied both in relation to physiology and to genetics. It is a convenient laboratory organism because it grows quickly, reproduces readily in culture and has basidiospores that germinate freely. A single basidiospore does not normally give rise to a mycelium bearing basidiocarps. Instead a branched mycelium of uninucleate cells without clamp-connections is established. This is known as a monokaryotic mycelium, or, more briefly, as a monokaryon. It produces an asexual stage in which tiny uninucleate spores (known as oidia or arthroconidia) are formed on short, erect hyphae in drops of slime. In nature these oidia are probably dispersed by insects and can reproduce the monokaryon.

When two compatible monokaryons meet, anastomoses occur. Nuclear migration from one to another follows and binucleate cells are established (Fig. 5.10). From these a dikaryotic mycelium (dikaryon) is formed made up of binucleate cells with clamp connections (clamps) at the septa. Basidiocarps are normally formed only on the dikaryon.

It is now necessary to consider how clamps are formed (Fig. 5.11A). A small, backwardly projecting lateral outgrowth occurs from the end cell of a hypha at some distance from its apex. Into this one of its two nuclei passes. Both then divide simultaneously and wall formation occurs between both pairs of daughter nuclei, the result being a binucleate end cell (at the base of

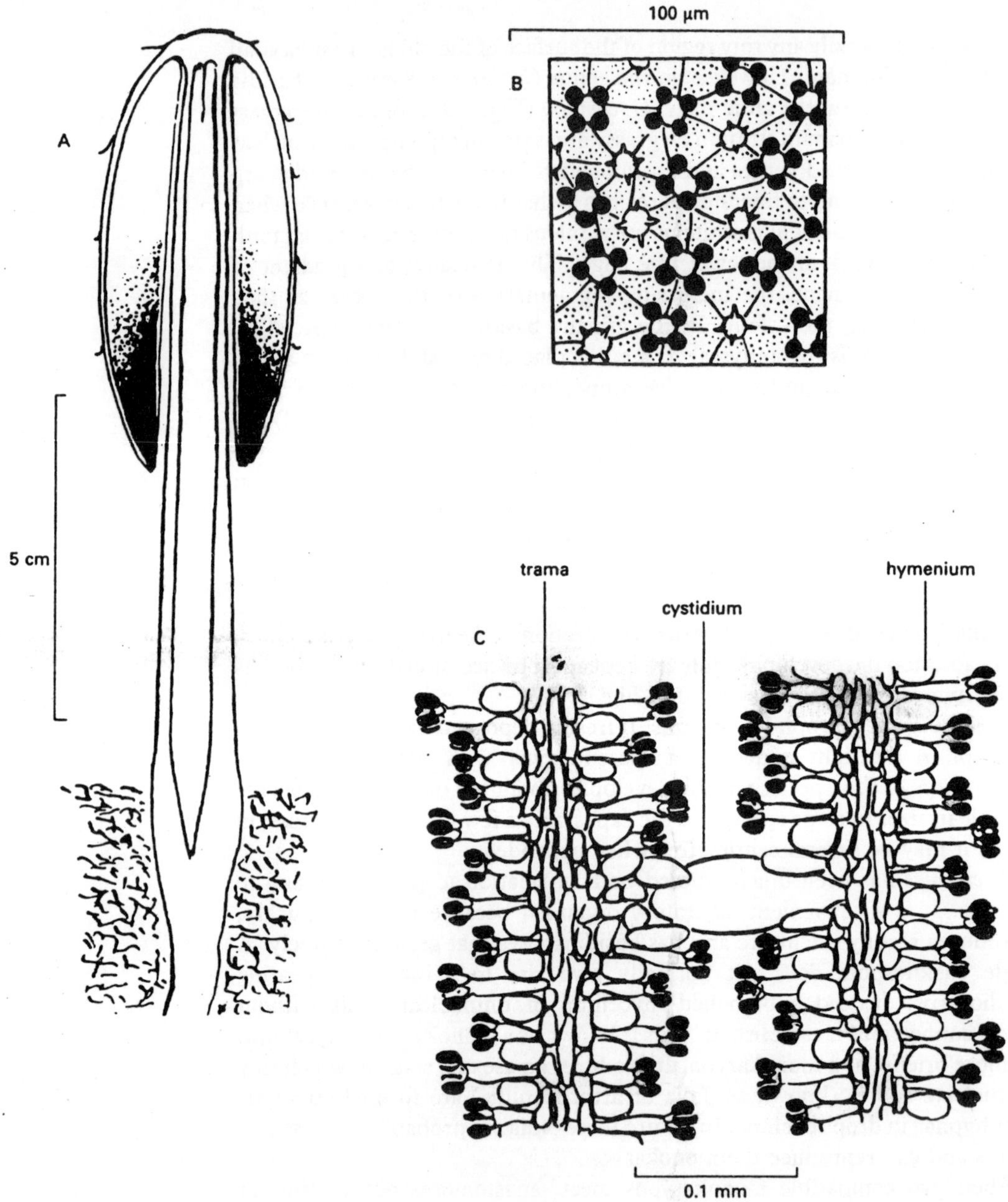

Fig. 5.9. *Coprinus comatus.* A, basidiocarp in longitudinal section; the gills are nearly vertical, each grading from white above through pink (dotted region) to black below. B, surface view of black region under high power; some basidia have just shot off their spores but the vacated sterigmata are showing; in others the black spores are about to be discharged; the basidia are set in a pavement of paraphyses (stippled). C, *Coprinus cinereus*, section through gills showing basidia of two lengths and paraphyses; also a cystidium keeping the gills apart. (C after Buller)

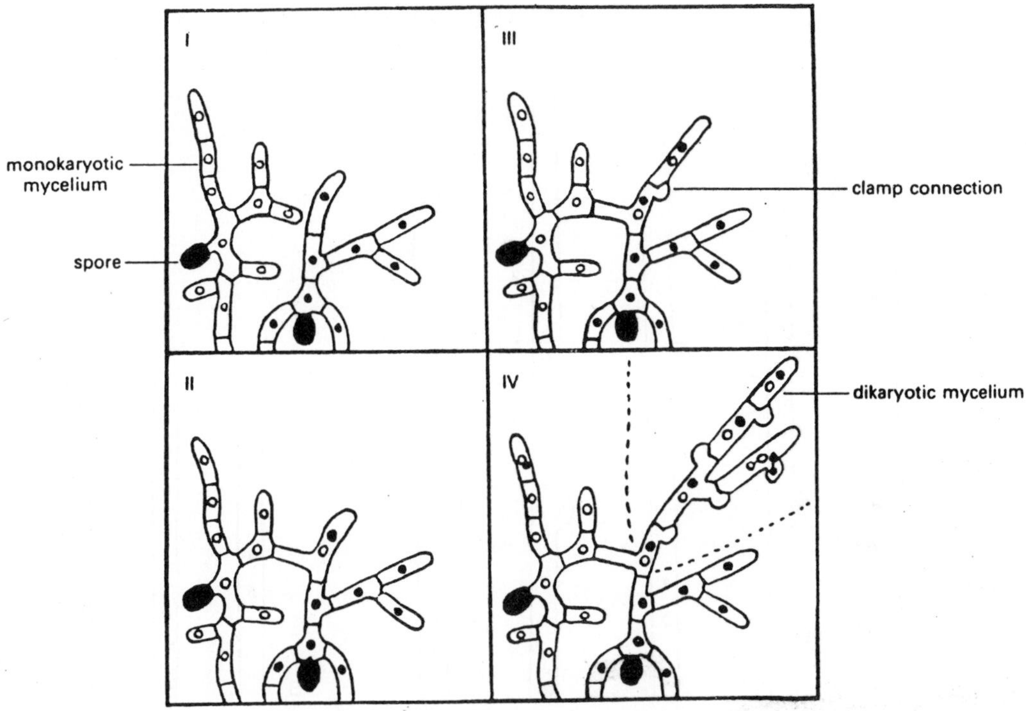

Fig. 5.10. *Coprinus cinereus.* I–IV, stages in the union of two compatible monokaryons with the formation of a dikaryon with clamp connections.

which is a tiny lateral uninucleate cell) and a uninucleate penultimate cell. The tip of the lateral cell now fuses with the penultimate cell and discharges its nucleus into that cell. At the same time the two nuclei of the terminal cell migrate to a middle position in preparation for the production of the next clamp. This kind of division ensures that, in each cell of the dikaryon, the two nuclei are of both types contributed by the monokaryons from which the dikaryon developed. Clearly if, in a narrow hypha, simultaneous division of the two nuclei occurred followed by wall formation to produce a binucleate condition (Fig 5.11B), the apical cell would immediately acquire two genetically identical nuclei. Clamp production is unique to Basidiomycotina, although it bears some resemblance to the process involved in repeated ascus formation at the tip of ascogenous hyphae (Fig. 4.1).

It should be noted that, although clamps are so characteristic of Basidiomycotina, there are some species in which they do not occur. They are not to be seen in the mycelium of *Agaricus* or in that of the honey fungus (*Armillaria mellea*). Further, in some species the development of clamps in the dikaryon depends on environmental conditions. Thus in *Coprinus micaceus* there are relatively few clamps if a culture is grown at 10°C, but abundant clamps are formed at 20°C. Also, in that species, the first-formed septum in a lateral branch from a hypha with clamps is always simple, and only later in its

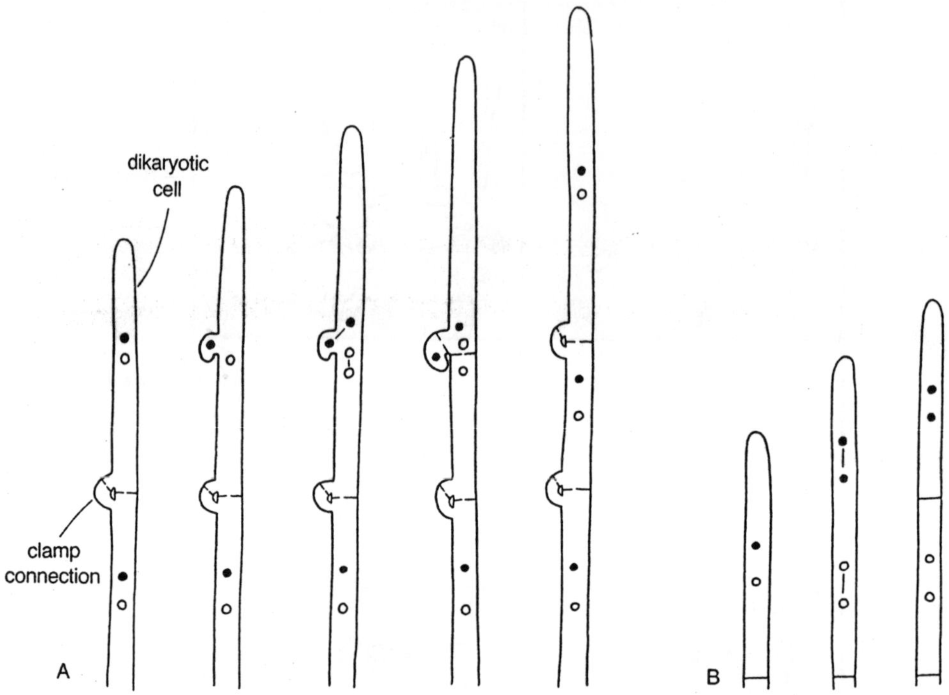

Fig. 5.11. A, diagram of nuclear and cell division resulting in clamp formation in a hymenomycete; nuclei of one mating-type shown white; the other black. B, what might happen if clamps were not formed.

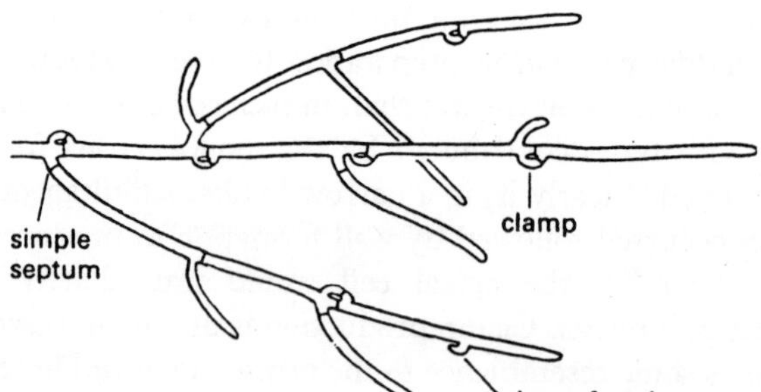

Fig. 5.12. *Coprinus micaceus.* Growing margin of a culture only diagrammatic to the extent that the individual cells are shown much shorter than they actually are. (Highly magnified)

growth do clamps form (Fig. 5.12). Where clamp formation fails, presumably some other mechanism maintains the dikaryotic condition.

In *Coprinus*, and other Basidiomycotina, each cross-wall is pierced by a central pore around which the cell wall is thickened to form a barrel-shaped

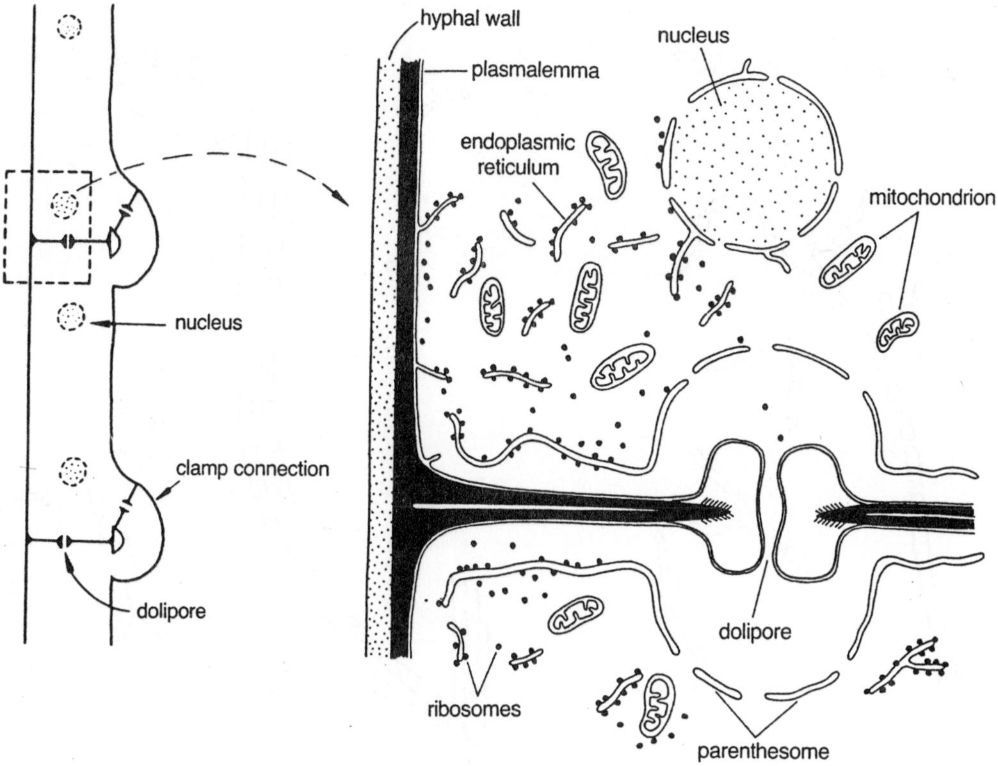

Fig. 5.13. Diagram of part of a dikaryotic hypha with one whole cell and two clamps. An indicated part is shown as it would appear under the electron-microscope.

structure known as a 'dolipore' (Fig. 5.13). Further, within the protoplasm the ends of the barrel are covered by perforated caps (parenthesomes) of endoplasmic reticulum. The dolipores seem to prevent the passage of nuclei from cell to cell, while perhaps allowing the movement of smaller organelles.

When compatible monokaryons meet, extensive nuclear migrations occur from one mycelium throughout the other. This is possible because the dolipores break down in advance of the migrating nuclei.

Reference has already been made to the fact that, in addition to producing basidiospores, *C. cinereus* also forms accessory asexual conidia. In many genera besides *Coprinus* such spores are formed, but mostly they are confined to the monokaryon. However, in *Flammulina velutipes*, the velvet-stalked mushroom, common in winter on dead elm (*Ulmus*), such spores are formed abundantly both on the monokaryotic and on the dikaryotic mycelium. These spores occur as simple or branched chains of dry, uninucleate spores formed, apparently, by fragmentation in terminal hyphae (Fig. 5.14). They are oidia or arthroconidia. Like the basidiospores of *Flammulina velutipes*, they germinate

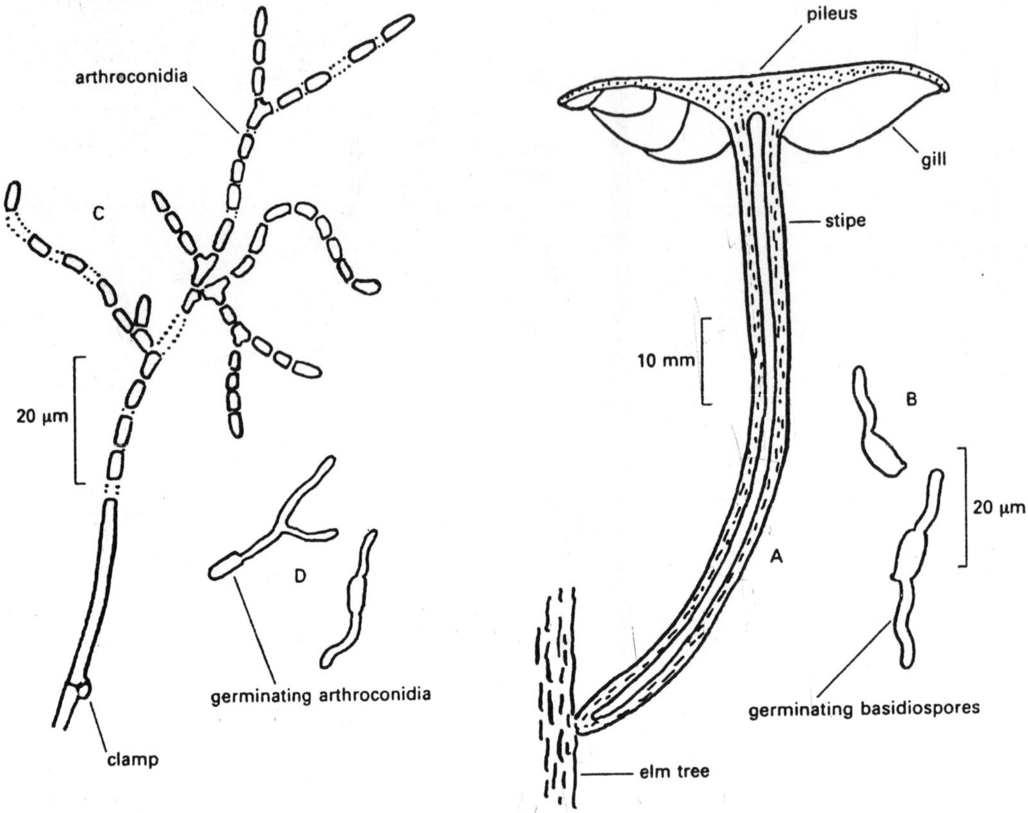

Fig. 5.14. *Flammulina velutipes.* A, basidiocarp in longitudinal section growing on elm. B, germinating basidiospores. C, chain of arthroconidia (oidia) produced from the clamped (dikaryotic) mycelium. D, arthroconidia from the dikaryon germinating to form a monokaryon.

readily and rapidly. Since the arthroconidia are uninucleate, each gives a monokaryon whether it comes from a monokaryon or from a dikaryon. Thus in the dikaryon, arthroconidia formation is a process of dedikaryotization.

The essential nature of Hymenomycetes has so far been illustrated by considering a few gill-bearing fungi. Attention must now be given to other types of basidiocarp which are to be found in Aphyllophorales (Fig. 5.2).

A considerable number of large and conspicuous fungi of woodlands belong to the family Polyporaceae in which the hymenium lines narrow vertical tubes or pores on the underside of the basidiocarp. In that family the texture of the basidiocarp is tough, tending to be leathery or even woody. Mostlv the basidiocarp takes the form of a bracket growing out from a living, or more usually dead, trunk or branch of a tree. These basidiocarps last longer than fleshy toadstools with an active life extending over several months. However, the majority endure only for a single season, but there are notable exceptions in the genera *Ganoderma* and *Fomes* in which the basidiocarps are perennial.

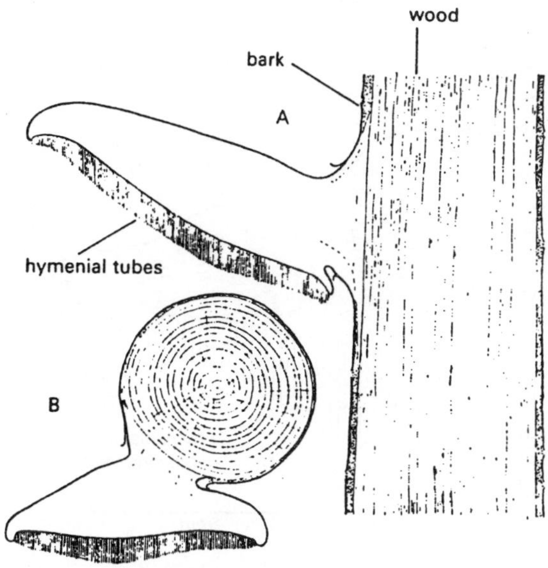

Fig. 5.15. *Piptoporus betulinus.* A, basidiocarp growing out of trunk of birch (*Betula*). B, another developing from a horizontal branch (× 2/3).

Piptoporus betulinus (Fig. 5.15) is a bracket polypore to be seen on dead standing and fallen trunks of birch (*Betula*) in almost any birchwood. The fungus is a wound parasite of weak trees, the basidiocarps being formed only after a tree is dead. The feeding mycelium is in the wood and the basidiocarp is produced as a lateral, semicircular bracket 20–40 cm across and 2–3 cm thick. Its more or less horizontal growth is a response to gravity. It is diagravitropic. At first the underside of the basidiocarp is smooth, but soon shallow pores are to be seen and these then grow downwards under the influence of gravity (positive gravitropism) to produce perfectly vertical tubes 10–15 mm long and 0.5 mm wide. The hymenium is just like that of an agaric and lines the tubes. The horizontal basidia shoot their spores into the cavities of the tubes. The spores then fall in the still air within the tubes and, emerging below the basidiocarp, are dispersed by wind. Basidiocarps in this fungus begin to form in August, are active during suitable weather throughout autumn and winter, finally decaying in the spring.

Most leathery and corky bracket fungi are drought-enduring fungi. During wet weather water is absorbed and spore discharge goes on. Under dry conditions water is lost, discharge ceases, and the basidiocarp remains alive but dormant until rain again restores it to activity.

We may now turn to a perennial form. *Ganoderma adspersum* is common in the form of thick, woody brackets, sometimes more than 50 cm across, rigidly and broadly attached to the trunks of dead and dying trees, especially beech (*Fagus*) (Plate 2c). This is probably the largest of all fungi, outmatching in weight, if not in volume, the giant puffball (*Calvatia gigantea*). In Fig. 5.16

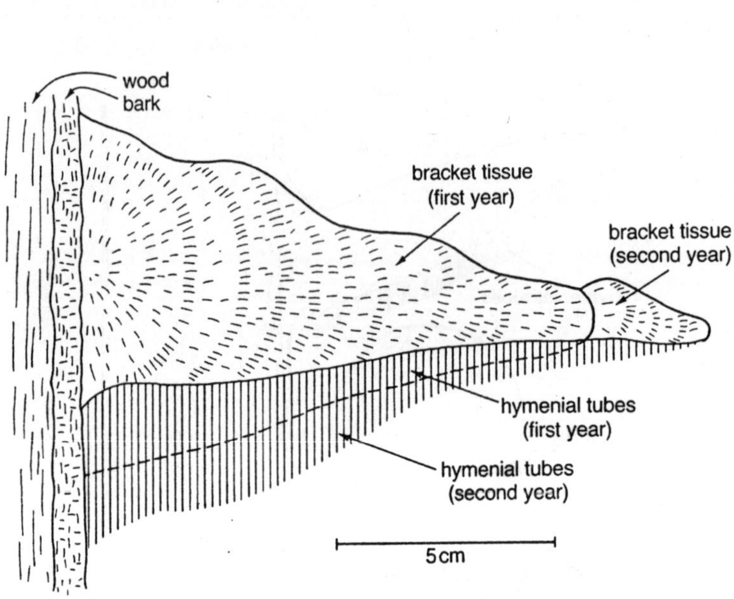

Fig. 5.16. *Ganoderma adspersum*. Vertical section of a small specimen on ash (*Fraxinus*); a line indicates the boundary between the growth of the hymenial tubes in their first and in their second year.

the structure of a small specimen in its second year is illustrated. The sterile 'cap' tissue grows out more or less horizontally. The hymenium-lined tubes are produced on the under surface. These are several centimetres long and extremely narrow (0.1–0.2 mm) (Fig.5.17). The tubes are absolutely vertical and the whole structure is so rigid that displacement cannot occur. In most other polypores the tubes are considerably wider, and there is some margin of safety. However, apparently because of its solid construction and being so broadly and firmly attached to a tree trunk, this margin can be reduced to practically nothing in *Ganoderma*. The tubes are little wider than the distance of spore discharge.

The hymenial tubes increase in length from May to September, but activity ceases during the cooler months, to be resumed in late spring. The cap tissue then increases around the circumference of the semicircular bracket with new tubes differentiating below this fresh tissue, and the tubes of the previous season recommence their downward growth. Structural weaknesses at the junction of one season's growth of tubes with that of the following season is such that, if a vertical incision is made in the cap tissue and the basidiocarp is broken apart, annual layers of tubes are easily distinguishable. A basidiocarp may continue to live for five to ten years, but only the last two or three layers of tubes have active hymenia. After a few years' service the older parts of the tubes become plugged with sterile hyphae and pass out of commission.

A large specimen of *Ganoderma adspersum* may liberate 20 000 000 spores a minute, and this fantastic rate may be maintained for the five months of the annual spore-fall period. If the space below a bracket is viewed on a bright day

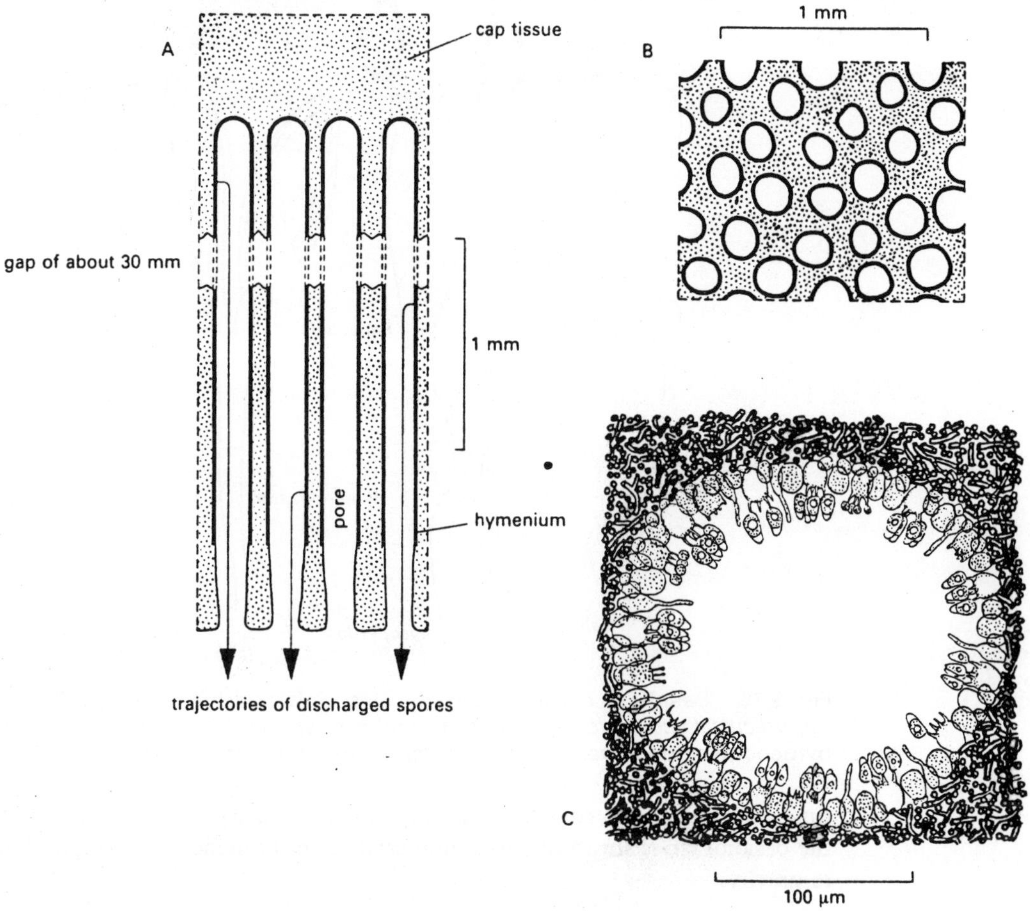

Fig. 5.17. *Ganoderma adspersum.* A, part of a vertical section of a basidiocarp. B, horizontal section through the pore region of a basidiocarp. C, details of a single pore.

in summer with the sun's rays slanting in the right direction, rust-coloured clouds of spores may be seen escaping steadily from the underside of the bracket.

Unlike most bracket fungi, *Ganoderma adspersum* can continue to discharge spores in the driest weather. This would seem in part to be due to the conduction of water through the hyphae from water in the wood, and in part to reduced water-loss from the hard, upper crust of the basidiocarp but also by the production of metabolic water from the breakdown of cellulose in the wood.

The same general principles govern the construction of the basidiocarps in other members of Aphyllophorales (Fig. 5.2). Thus in *Hydnum* teeth or spines covered by hymenium occur on the underside of the cap. These, like gills in an agaric, are positively gravitropic. In *Clavaria* and its allies, known as coral

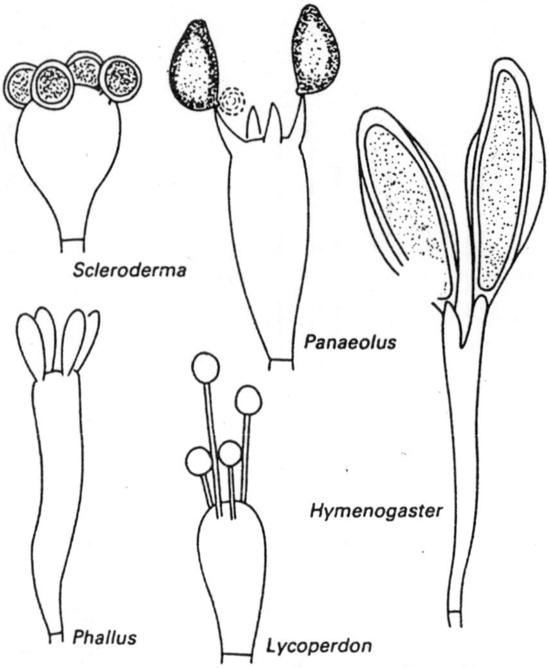

Fig. 5.18. Basidia in some Gasteromycetes. *Scleroderma citrinum*; *Phallus impudicus*; *Lycoperdon gemmatum*; and *Hymenogaster citrinus* and in the hymenomycete *Panaeolus campanulatus*. (All highly magnified)

fungi, the hymenium covers a simple club which is negatively gravitropic, or the basidiocarp is shrub-like with numerous erect branches each responding to gravity.

GASTEROMYCETES

Included in Gasteromycetes are puff-balls, earth stars, bird's nest fungi and some underground types. None is known to be poisonous, nor do any cause disease in higher plants. It is a harmless assemblage of fungi. The spore-producing surfaces (hymenia) are not exposed at maturity, and the basidium is not a spore-gun. The basidiospores may be sessile on the basidium or borne on sterigmata, but then the elegant asymmetrical poising of the spore on its sterigma, so characteristic of Hymenomycetes, is not to be seen (Fig. 5.18). Many mycologists consider that Gasteromycetes have been derived, along a number of separate lines of descent, from hymenomycete ancestors. Having lost the hymenomycete equipment of spore liberation, Gasteromycetes appear to have developed new methods of dispersal along original lines.

In a puff-ball (*Lycoperdon*) the developing basidiocarp, at a fairly advanced stage of development, has the appearance shown in Fig. 5.19. In the fertile region (gleba) there are numerous irregular chambers lined by basidia and separated by tramal tissue. The basidia, which are not nearly so regularly

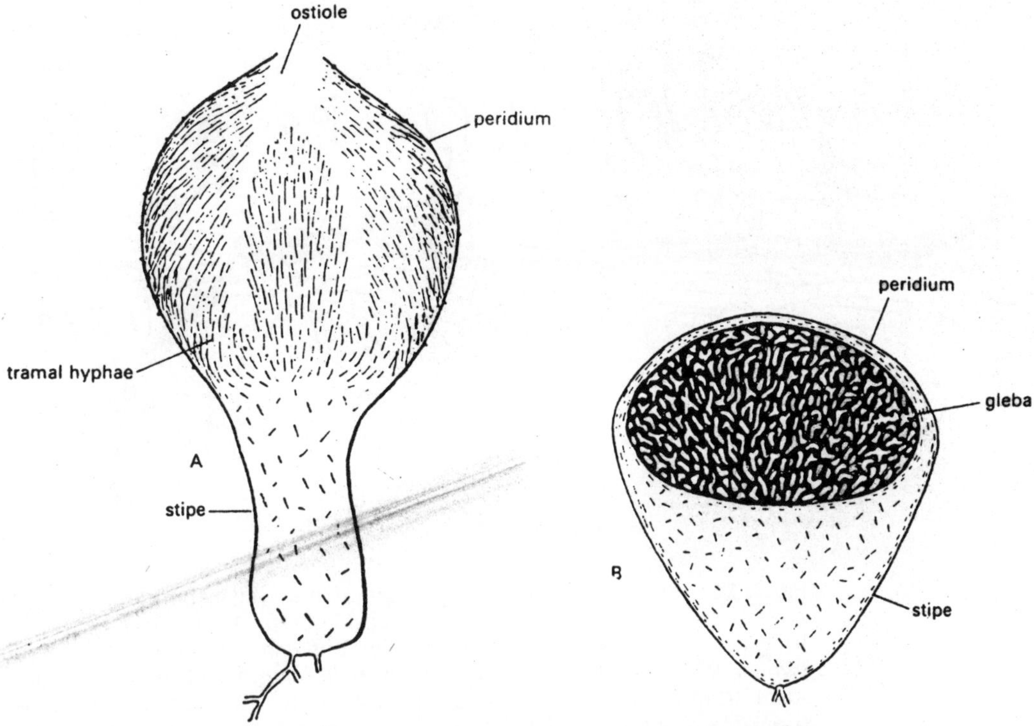

Fig. 5.19. *Lycoperdon*. A, *L. gemmatum*: longitudinal section of a basidiocarp, life-size. B, *Lycoperdon* sp.: longitudinal section of a young specimen in which the glebal chambers lined with hymenium are shown white; twice natural size.

arranged as in an hymenomycete, have slender, cylindrical sterigmata each capped symmetrically by a spherical basidiospore. The tramal tissue consists partly of thin-walled hyphae and partly of hyphae with very thick walls (Fig. 5.20). As the basidiocarp matures, the basidia break down leaving the spores free. Further, the tramal tissue breaks down with the exception of the thick-walled hyphae. These persist as a springy mass of dry threads, rather like cotton-wool, with the basidiospores intermixed as a dry powder. The wall (peridium) of the basidiocarp dries to form a papery waterproof covering. This ruptures at the top of the basidiocarp to form a definite mouth (ostiole). Wind blowing across this no doubt sucks out spores, but it seems that essentially the puff-ball is a bellows operated by large rain-drops or dripping water from overhead trees. When a drop strikes the peridium, this is momentarily depressed. The immediate result is a puff of air through the ostiole producing a conspicuous cloud of spores. The mechanism can readily be demonstrated by allowing large drops to fall singly from a height on to a ripe puff-ball.

The number of spores in a puff-ball is enormous. It has been estimated that a single basidiocarp of the giant puff-ball (*Calvatia gigantea*), which may be up to 40 cm across, produces seven million million basidiospores.

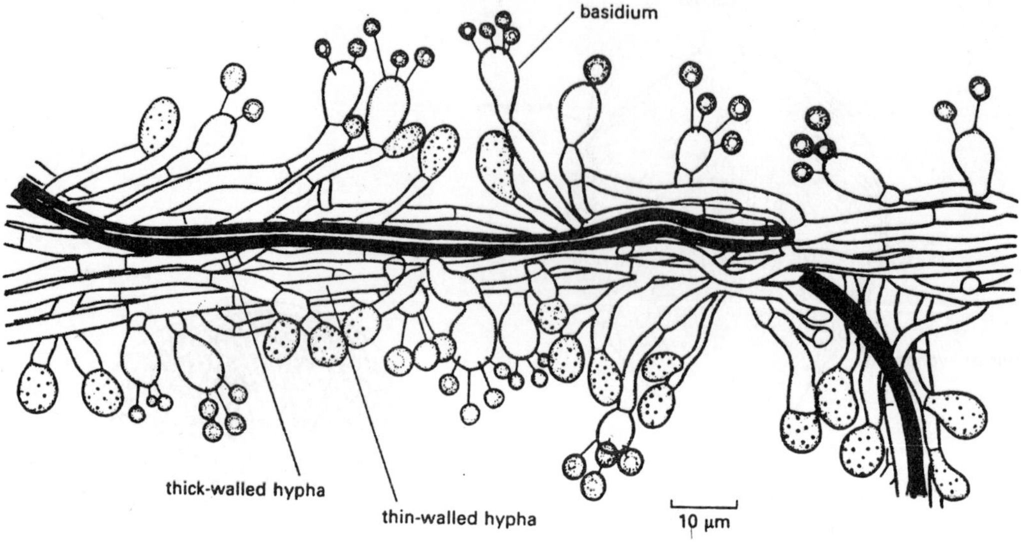

Fig. 5.20. *Lycoperdon pyriforme*. Part of the wall separating two glebal chambers in the gleba of a young basidiocarp. Above and below is an irregular layer (hymenium) of basidia; between are thin-walled hyphae (which will break down) and very thick-walled hyphae. These will persist to form a mass (like cotton-wool) in which the dry spores (set free by breakdown of the basidia) occur. (After Webster).

The stink-horn, *Phallus impudicus* (Fig. 5.21), is common in woods from June to October, and it is rare to walk through a wood in that period without from time to time encountering its unpleasant smell, like that of a cess-pit. The feeding mycelium is in buried wood from which branched, white hyphal strands are formed. These may pass through the soil for many centimetres before basidiocarps are produced. The basidiocarp develops just below the leaf-litter in a wood, and when nearly mature resembles, in size and shape, a hen's egg, but a soft one. The outer part (peridium) consists of a skin, a thick middle layer of jelly, and a membraneous innermost layer (Fig. 5.22). The organization of the gleba within is rather like that of *Lycoperdon*. However, the basidia break down to give, not a dry, but a slimy mass of spores in a sugary matrix. When the basidiocarp is ripe, the stipe elongates with dramatic suddenness, growing to full length in 2–3 hours and carrying the cap, covered with its slimy coating of spores, to a height of 12–15 cm. For this spectacular growth water is required, and this is absorbed from the jelly of the peridium which represents a water reserve slowly built up, but rapidly utilized. Shortly after the cap is raised into the air, it begins to emit its characteristic odour. This attracts a variety of large flies, and soon the cap is crowded with a greedy mass of milling insects. In a few hours the slime is completely devoured, leaving a white cap covered with a honeycomb of shallow pits. It seems that the flies disperse the spores which adhere to their legs and also pass, apparently uninjured, through the alimentary tract. There is, however, a serious gap in our knowledge of the biology of the stink-horn and its allies. No

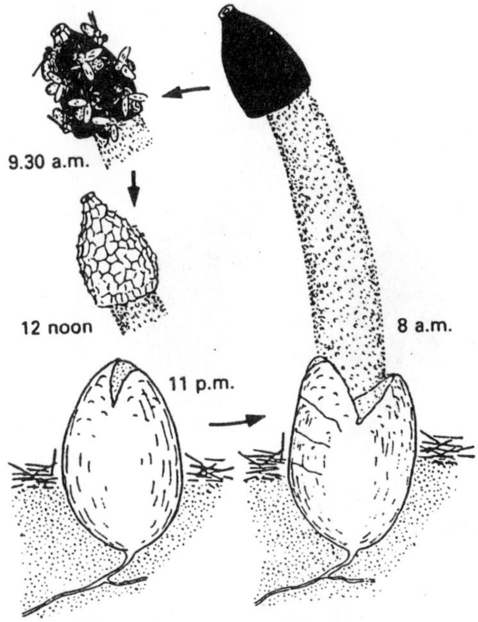

Fig. 5.21. *Phallus impudicus.* At 11 p.m. the basidiocarp is at the 'egg' stage and the outer, papery layer of the peridium has torn exposing the jelly below. Next morning at 8 a.m. the stalk has elongated carrying up the cap bearing the spore slime. By 9.30 a.m. the slime is giving out a strong smell and has attracted flies. By noon all the spore slime has been removed by the flies.

one has, apparently, yet succeeded in germinating the spores whether taken directly from the fungus, or after passage through an insect.

Gasteromycetes also include the beautiful bird's nest fungi (*Cyathus* and *Crucibulum*) which have been the subject of intense study. Most species are found on rotting wood or sticks, but *Cyathus stercoreus* occurs on herbivore dung. The mature basidiocarp is an open vase or cup, about 10 mm in diameter, in which a number of hard discoid bodies (peridiola) lie like eggs in a nest. In the developing basidiocarp, the gleba is divided into these separate portions each surrounded by a firm wall of interwoven hyphae attached by a delicate hyphal cord (funiculus) to the inner wall of the cup. Large rain-drops (3–5 mm across) falling into the cup are broken up into droplets which are reflected out of the cup. The larger of these may pick up peridiola and carry them to a distance of up to a metre (Fig. 5.23). The peridiolum takes the funiculus with it, for this is very loosely fixed to the cup. Further, the funiculus is sticky and cements the peridiolum to any object which it strikes. There is little doubt that the basidiocarp in *Cyathus* and *Crucibulum* is a specialized structure that has developed in relation to splash dispersal.

On a suitable substratum, a peridiolum may give rise to a new mycelium simply by growth from the hyphae forming its wall without the basidiospores being involved. However, in the coprophilous *Cyathus stercoreus* the peridiola may be eaten with grass, and the dispersal of basidiospores may then occur as in other coprophilous fungi.

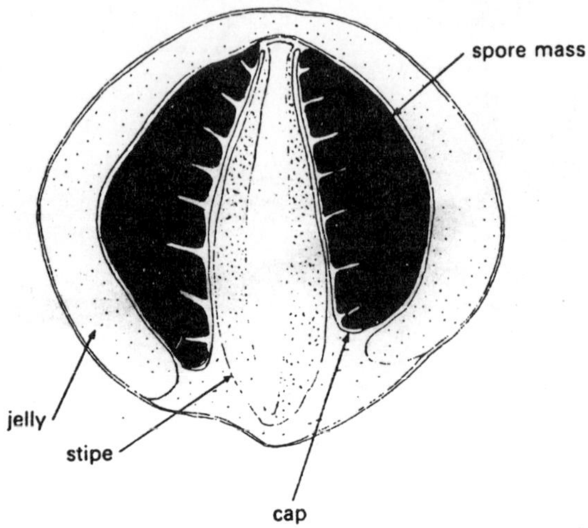

Fig. 5.22. *Phallus impudicus.* 'Egg' in longitudinal section (life-size); the gleba consisting largely of spores is shown black.

In *Sphaerobolus* (Fig. 5.23C) an even more spectacular mechanism is found. This fungus is fairly common on rotten wood and on old pads of cow-dung. It is a tiny Gasteromycete and the basidiocarps, which usually grow crowded together, are only 1–2 mm across. When young each is spherical and about the size and colour of a swollen mustard seed. The mature basidiocarp splits open and is seen to be in the form of two little cups of peridial tissue one inside the other and in contact only at the rims. Inside the inner cup is a soft glebal mass about 1 mm in diameter lying loosely in a lubricating fluid. The inner cup is composed of highly turgid tissue in which strains develop. As a result it suddenly turns inside out catapulting the glebal mass to a distance of several metres. It is by far the most powerful piece of fungal gunnery, outclassing *Pilobolus* (p. 30). As with the peridiola of bird's nest fungi, the discharged glebal mass of *Sphaerobolus* can germinate to produce a new mycelium without the basidiospores being involved. In addition to the basidiospores there are some large dikaryotic cells (gemmae) in the glebal mass, and it is from these, rather than from the basidiospores, that mycelium normally develops. Reference has been made to *Sphaerobolus* on cow-dung. Where this occurs it is likely that the glebal-masses, which stick firmly to herbage, have been eaten by cows and that dispersal has followed the usual course for coprophilous fungi.

Not only is *Sphaerobolus* quite common, but, unlike most other Gasteromycetes, it grows and reproduces freely in pure culture on agar.

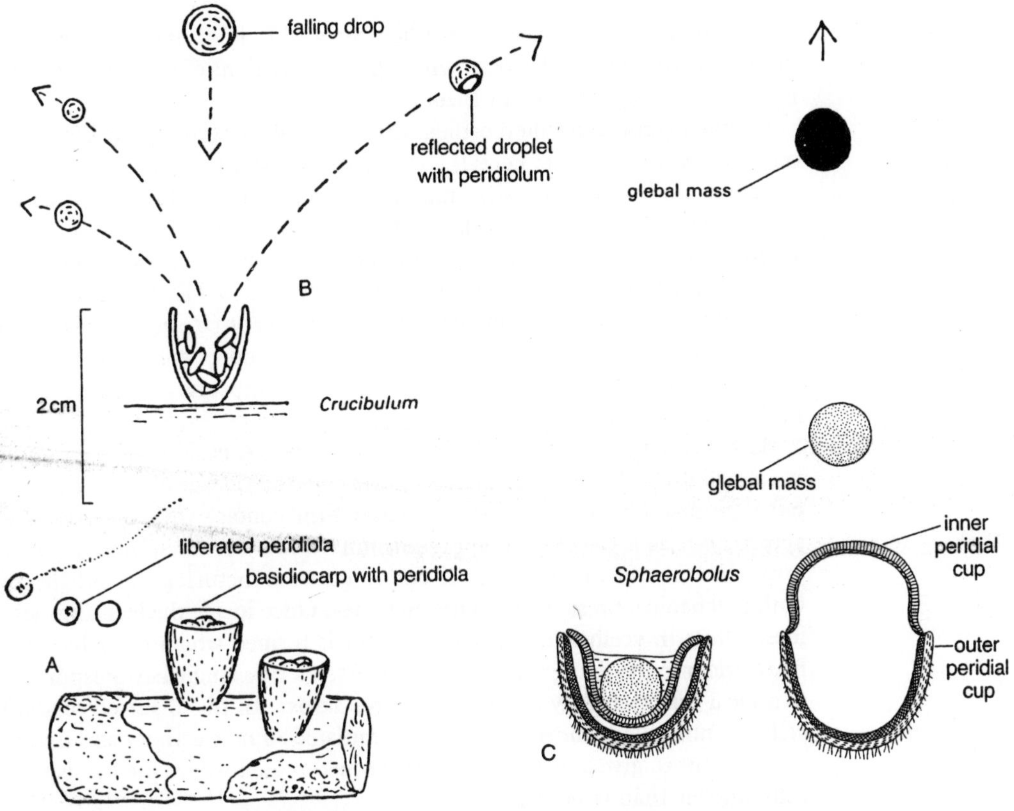

Fig. 5.23. A and B *Crucibulum vulgare*; A, two open basidiocarps growing on the wood of a partially decorticated stick; three liberated peridiola are also shown, one with its funiculus extended. B, diagram of splash dispersal showing falling drop and subsequent reflection of this; one of the reflected droplets bearing a peridiolum. C, *Sphaerobolus stellatus*: left, open basidiocarp in longitudinal section ready to discharge its glebal mass and (right) at moment of discharge.

Certain Gasteromycetes have subterranean basidiocarps (e.g. *Hymenogaster*). As in *Tuber* in the Ascomycotina, rodents grub these up, eat them and apparently disperse the spores through the gut.

TELIOMYCETES

Uredinales

In the Teliomycetes basidiocarps are lacking and all are specialized plant parasites. The Uredinales, the rusts, constitute the most important order with over 150 genera and about 6000 species. They occur in nature only as biotrophic parasites of living plants. However, a few have now been grown with considerable difficulty in pure culture or nutrient agar.

Rusts will be discussed mainly by considering *Puccinia graminis*. This is the 'black stem rust' of wheat and other cereals. It is of enormous economic importance in most of the great wheat-growing regions of the world, although it is of minor significance in Britain.

Puccinia graminis is heteroecious, that is to say it has part of its life-cycle on one host (wheat or other cereal) and the rest on another (barberry) of a very different kind. Many parasitic animals are heteroecious, but, so far as fungi are concerned, the heteroecious condition is found only in rusts, although a number of these are autoecious, completing their life-cycle on a single host.

There are five types of spore in *P. graminis* (Fig. 5.24). Pycniospores and aeciospores are produced on the common barberry (*Berberis vulgaris*), and urediniospores and teliospores on wheat. Basidiospores are borne on basidia which arise on germination of the teliospore. We must now consider the life-cycle in considerable detail.

Barberry is infected in spring by basidiospores. A basidiospore germinates in a small drop of water, the infection drop, on the surface of the leaf, putting out a germ-tube the tip of which makes firm contact with the cuticle and spreads out as a sucker-like appressorium. While the circumference of this grips, like a limpet on a rock, the centre grows puncturing the cuticle using both mechanical pressure and lytic enzymes. Once inside the leaf a branched, intercellular mycelium of uninucleate cells is formed which feeds by putting haustoria into the living cells of the host. Attack by a single basidiospore leads to a local infection only a few millimetres across. In this diseased region the cells are not killed. Indeed, they are stimulated to new activity and divide to some extent to give a distorted hypertrophied region where the leaf is several cells thicker than is normal.

Infection of a leaf by a solitary basidiospore thus forms a diseased spot, and in due course on the upper surface of the leaf a group of flask-like pycnia is produced. Within the pycnium the pycniospores are formed. Its interior is lined by a palisade of long cells each producing minute, uninucleate pycniospores from its apex. These, suspended in a sugary liquid, ooze out through a narrow mouth (ostiole) (Fig. 5.25A). Around this is a fringe of stiff hairs which support the spore-containing droplet of pycnial nectar capping the pycnium. Growing into this droplet are also a few long flexuous hyphae. Further, in the diseased area of the leaf are little knots or nests of fungal hyphae embedded in the leaf tissue towards the lower surface. These are the proto-aecia, and, if the single-basidiospore infection is protected from outside contamination, they develop no further. If, however, the leaves are freely exposed so that visiting insects may bring in pycnial nectar from neighbouring infections, the proto-aecia soon develop into aecia.

Puccinia graminis and most other rusts are heterothallic. There are 'plus' and 'minus' basidiospores each giving rise to apparently identical infections. If pycnial nectar from a 'plus' infection is mixed with that of a 'minus' one on the leaf, it seems that the 'plus' pycniospores fuse with the 'minus' flexuous hyphae. What happens thereafter is largely speculative, but it seems probable that 'plus' nuclei pass into the 'minus' mycelium and migrate, dividing as they go, so that the monokaryotic mycelium in the diseased leaf is converted into a dikaryotic one as in the Buller phenomenon (see p. 142). The pycniospores do

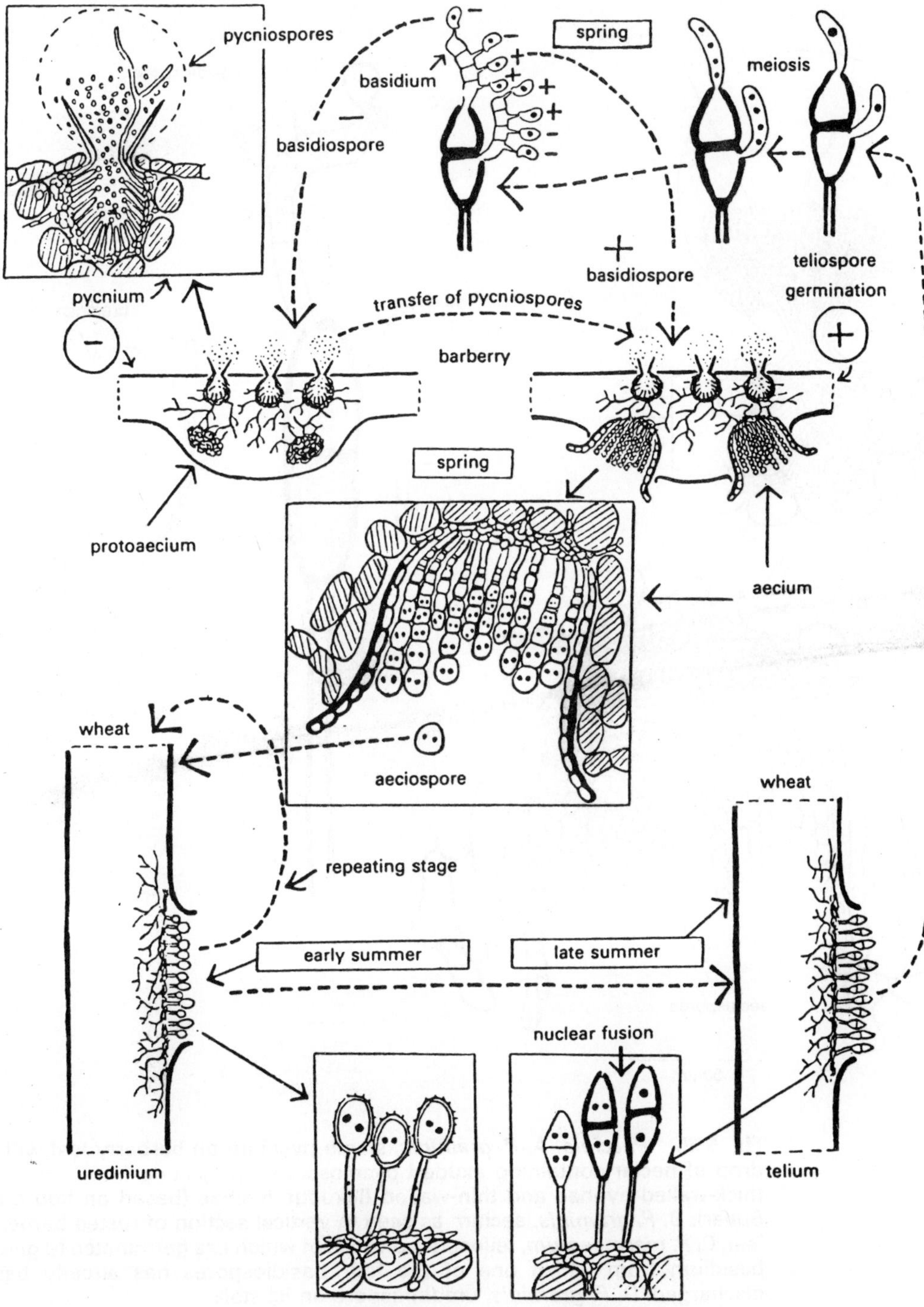

Fig. 5.24. *Puccinia graminis.* Diagram of the life-cycle. Dashed lines show the course of the cycle. Inset frames show details of some of the stages.

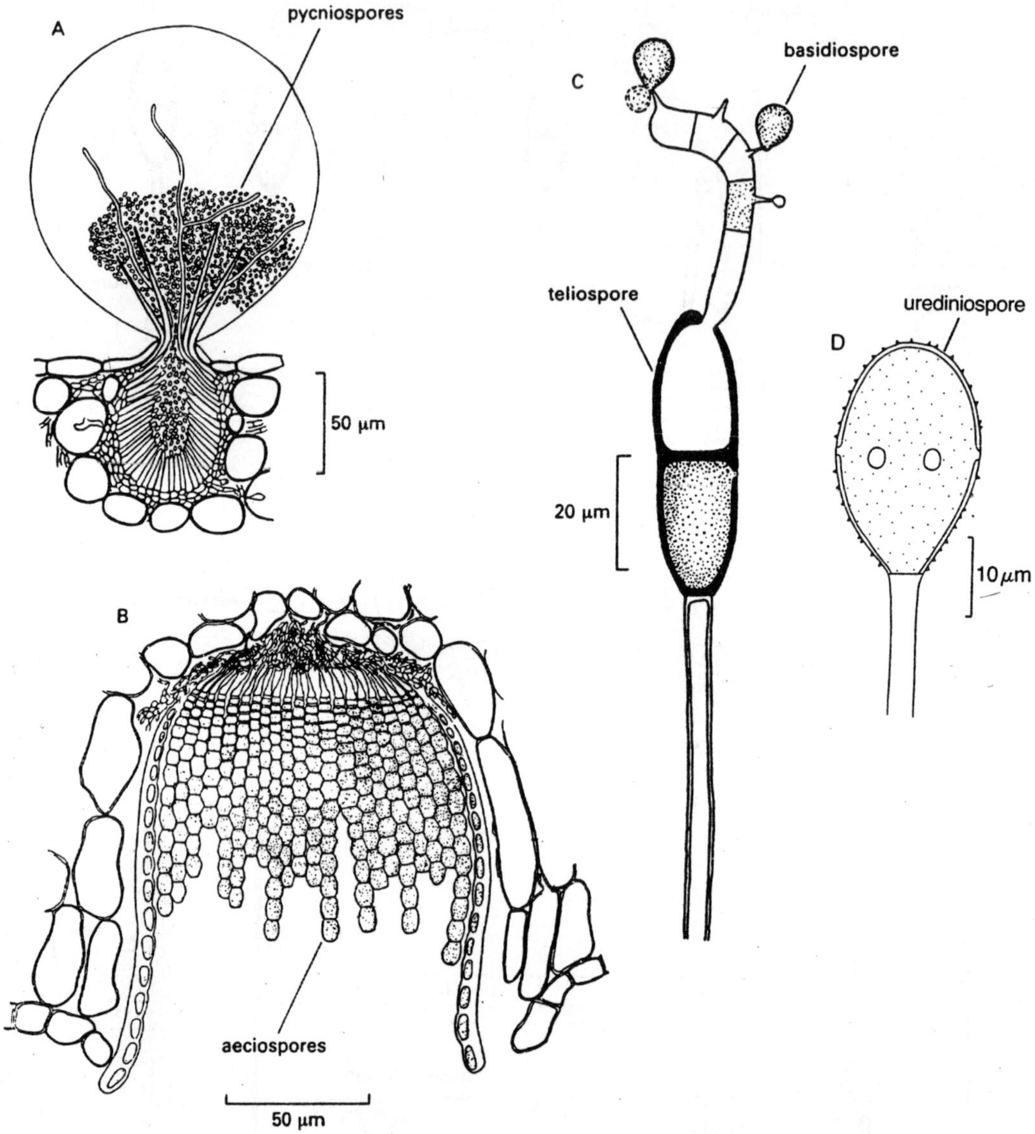

Fig. 5.25. *Puccinia*. A, *P. graminis*, single pycnium on barberry leaf with a drop of nectar containing exuded pycniospores; projecting into it are stiff thick-walled hyphae and thin-walled flexuous hyphae (based on figure by Buller). B, *P. graminis*, aecium as seen in vertical section of rusted barberry leaf. C, *P. malvacearum*, teliospore one cell of which has germinated to give a basidium from which one of the four basidiospores has already been discharged. D, *P. graminis*, urediniospore on its stalk.

not germinate by germ tubes. They are simply transporters of genes injecting their nuclei into a mycelium of opposite mating-type.

Aecia (Fig. 5.25B) are cup-shaped, and, since in most rusts they occur in small groups, this stage in the life-cycle is often called the 'cluster-cup' stage. The wall (peridium) of the aecium consists of thick-walled cells. At the base of the cup is a palisade of binucleate cells, each apparently with one 'plus' and one 'minus' nucleus. These basal cells produce growing chains consisting of binucleate aeciospores alternating with thin-walled separating cells which soon break down.

The aeciospores are violently discharged. They are turgid cells constrained within the aecium to a polyhedral form, but tending to round off. Strains are thus set up which are suddenly relieved, with the result that individual spores and groups of spores bounce out of the aecium and are carried away on the wind. Aeciospores are unable to infect barberry, but can attack wheat, or other members of the grass family, according to the strain of rust involved.

The orange–yellow aeciospore, which has quite a thick wall and contains lipid droplets, germinates in an infection-drop by one or several germ tubes. These enter the blade or base of the leaf by way of stomata. Within the host a localized, intercellular mycelium of binucleate cells is formed which feeds by haustoria inserted into the living cells. In due course in the local infected region a uredinium is produced. In its formation the mycelium ramifies richly below the epidermis of the leaf and from this bed of fungal tissue urediniospores are developed, each on an erect stalk-cell. As a result of the development of the urediniospores, the epidermis of the host is ruptured, exposing the reddish-brown mass of spores. It is at this stage, when the wheat plant is speckled with uredinia, that the characteristic rusted appearance is evident. Each urediniospore is binucleate, contains a lipid reserve and has a thickish, ornamented wall (Fig. 5.25D). The urediniospores are not violently discharged. Both aeciospores and urediniospores, unlike the thin-walled basidiospores with their meagre reserves, are spores that can be dispersed for long distances through the air without losing the capacity to germinate.

The urediniospores have the same capacity to infect as the aeciospores, being able to attack wheat, but not barberry. Further, their mode of entry and the result of infection is just the same; each produces a localized dikaryotic mycelium. Thus the urediniospore represents a repeating stage in the life-cycle. This is the epidemic phase in an attack by rust when the disease can spread rapidly through the wheat crop.

An infected plant may bear many uredinia and, although the host cells are not directly killed, great damage may be done to the plant as a whole. Its capacity to photosynthesize is reduced and its growth is checked. Further, uncontrolled water loss through the uredinia may lead to permanent wilting and shrivelling of the leaves under dry conditions.

The dikaryotic phase on wheat produces urediniospores during most of the summer, but later teliospores appear mixed with the urediniospores. Finally only telia are formed. These are black elongated areas on the stem and leaf bases in contrast to the rusty uredinia. The development of the telium and the arrangement of the spores within it are basically the same as in a uredinium.

However, the teliospore, borne on a unicellular stalk, is two-celled and its wall is very dark brown, smooth and thick except for a thin-walled germ pore in each cell (Plate 1E). The teliospore is not dispersive. It is a resting-spore. It stays on its stalk which itself remains attached to the straw when the wheat plant dies. The teliospore is not capable of immediate germination, but remains inactive over winter and is not ready to germinate until late spring.

At first the two cells of the teliospore are binucleate. However, as the spore ripens the two nuclei in each cell fuse. It seems clear that the association of 'plus' and 'minus' nuclei which began in the proto-aecium is completed by fusion in the teliospore. Throughout the prolonged dikaryotic stage each cell of the mycelium has, apparently, two nuclei of opposite mating-type.

Under appropriate conditions in spring, each cell of the teliospore forms a short, curved hypha, the young basidium, into which the diploid nucleus passes. This then undergoes meiosis, and the basidium becomes septate to produce four cells each with a haploid nucleus. Each cell now forms a single basidiospore borne asymmetrically on its sterigma (Fig. 5.25C). It is discharged as is the basidiospore in Hymenomycetes. It is a thin-walled spore incapable of infecting wheat, but can attack barberry in the manner already described.

It should be emphasized that of the five types of spore in the life-cycle of *P. graminis*, three (aeciospores, urediniospores and basidiospores) are dispersed by wind, pycniospores are spread by insects and teliospores are not dispersed at all.

Although many rusts are heteroecious with a striking contrast between the two hosts in the life-cycle, a large number are autoecious. Thus blackberry rust has all its stages on *Rubus fruticosus*. Many rusts have shortened cycles, with one or more types of spore left out. Hollyhock rust, *Puccinia malvacearum*, has only teliospores and basidiospores. Further, in this common rust the teliospore germinates without a period of rest. Although most rusts cause only local infection, a few are systemic with the mycelium invading the whole shoot. This is so in *Puccinia punctiformis*, thistle rust. In spring infected shoots of the common thistle (*Cirsium arvense*) are evident by their paler colour and spindly growth. At an early stage the undersides of the leaves of affected plants are covered with pale-yellow pycnia, and the whole shoot has a sweet smell. It is claimed that the same scent can be detected in the pycnial stage of *Puccinia graminis*. Like the nectar associated with the pycnium, this scent is, no doubt, associated with insect dispersal. In thistle rust the pycnial stage is succeeded by the chocolate-coloured uredinial stage and the sweet scent is lost. In this particular rust the aecial stage is omitted from the life-cycle.

Many years ago the Swedish mycologist Eriksson found that different grasses and cereals reacted in a different way to any given strain of *Puccinia graminis*. These are called 'special forms' and the infraspecific category 'formae speciales' (abbreviated 'f.sp.') has been designated for taxa characterized from a physiological standpoint but scarcely or not at all from a morphological one. Six are recognized in *Puccinia graminis*. For example *P. graminis* f.sp. *tritici* infects wheat, barley and some wild grasses but not oats; *P. graminis* f.sp. *avenae* infects oats and a few wild grasses, but not wheat and barley. Eriksson thought that the 'special forms' were morphologically identical. Later workers have shown, however, that statistically they can be distinguished on the basis of urediniospore size, and some now regard them as distinct varieties. All have teliospores producing basidiospores capable of infecting *Berberis vulgaris*.

Within a single forma speciale further physiological specialization may be found. If, for example, urediniospores of *P. graminis* f.sp. *tritici* are inoculated onto a range of wheat cultivars, these differ in their response. Some may be resistant, some highly susceptible and others may be intermediate in their response. Such differences are attributed to there being physiological races within *P. graminis* f.sp. *tritici*. Since the determination of which physiologic races are prevalent has become an important aspect of disease control in cereal rusts, the procedure for their recognition is of interest. If a particular race is tested simultaneously onto a standard set of differential wheat cultivars, its infectivity can be quantified and the race to which it belongs identified. The race is given a number: *Puccinia graminis tritici 138*.

With the knowledge of what races of rust are prevalent in a particular country, the breeder can develop varieties of wheat resistant to them. However, the situation may change. Races previously abundant may become less common, and rarer ones may increase in importance. The breeder must then supply new varieties to meet this changed situation. Apart from uncommon known races becoming abundant, new ones may arise either by mutation or by genetic recombination. Barberry is not only an operational base from which the rust may launch its attack on wheat in spring; it is also a meeting ground where different races may come together, cross and produce new combinations.

The existence of distinct physiologic races is not a phenomenon limited to rusts. It is a characteristic feature of specialized parasites generally, and in fungi it is well known in both powdery and downy mildews.

In a heteroecious rust such as *P. graminis*, it might be expected that control could be achieved by destroying the alternate host (barberry), and in certain regions this is effective. In Britain the comparative rarity of barberry, coupled with the fact that the British winter is normally too cold for urediniospore survival, ensures that black rust is not an important disease. In the great wheat-growing areas of Canada the winter, as in Britain, is too cold for urediniospores to survive, and it might have been expected that barberry elimination would effect complete control. However, this is not so, for urediniospores, blown from many hundreds of miles further south where the rust develops earlier, may initiate infection in the spring. In Australia black rust is important in spite of the rarity of barberry, for there the urediniospores are not destroyed in the milder winter. It is to be noted that rusts differ from one species to another in the ability of their urediniospores to withstand cold. Thus *Puccinia striiformis*, yellow-rust of wheat, which probably reduces yield in Britain by 5–15%, has urediniospores that can endure temperatures below freezing point. *P. striiformis* has no known alternate host and appears to reproduce entirely by urediniospores.

Although in such a rust as *P. graminis*, with a repeating (uredinial) stage, control by elimination of the aecial host is not always effective, in rusts without such a stage this kind of control is complete. In *Gymnosporangium juniperi virginianae*, a heteroecious species attacking apple and red cedar in North America, there is no urediniospore stage, and orchards can be completely protected by removal of nearby red cedar (*Juniperus virginiana*).

The genera of Uredinales are distinguished mainly on the basis of the teliospore. It is one-celled in *Uromyces*, two-celled in *Puccinia* and three–eight-celled in *Phragmidium*.

Ustilaginales

Ustilaginales or smuts comprise the other, much smaller, group of plant pathogens usually classified in the Basidiomycotina. Although in nature they are known only as parasites, unlike the rusts they are not obligate biotrophic ones. Many have been cultured through their whole life-cycle. They are called smuts because at the most, and usually only, conspicuous stage of their life-cycle, they form dry, dusty, smutty or sooty masses of spores, teliospores. They inhabit stems, leaves and floral parts and some even cause galls or tumours on roots. In many hosts the fungus is systemic forming a fine intercellular mycelium which aggregates in specific parts of the host where teliospores are formed.

Loose smut of oat, *Ustilago avenae*, may be especially mentioned. This disease is not serious for it is easily controlled by appropriate treatment of the seed, and is not now often to be seen in a field of oats. However, it is also abundant on the common oat-grass (*Arrhenatherum elatius*). Smutted plants are conspicuous since the ovaries and some of the associated structures are reduced to masses of black teliospores, sometimes referred to as chlamydospores or brand spores.

In oats, smutted inflorescences (ears) emerge slightly before the healthy ones. The spores are carried by the wind to impact onto the stigmas or ovary walls of healthy plants at the time of antithesis. The dikaryon, when formed, grows out and penetrates the host often through the pericarp wall. The mycelium lies dormant there until the seed is sown the following season. As the seed germinates, the mycelium grows and keeps just behind the growing point until it reaches the developing ear. It amasses there and fills the developing grain with teliospores.

In the production of teliospores the dikaryotic mycelium of the fungus aggregates in all the ovaries of the infected plant. Many of the cells of this mycelium then swell to form teliospores with fairly thick walls. As each teliospore matures, its two nuclei fuse.

On germination a single germ tube (the young basidium) is formed, and into it the diploid nucleus passes. This then undergoes meiosis. A four-celled basidium is normally produced, and from each cell of this a uninucleate basidiospore is budded (Fig. 5.26). It is not discharged. Of the four spores of a basidium, two are 'plus' and two 'minus' as in rusts. A basidiospore, with or without detachment, readily buds like a yeast. If a single basidiospore is transferred to nutrient agar, a yeast-like colony is formed and the fungus can

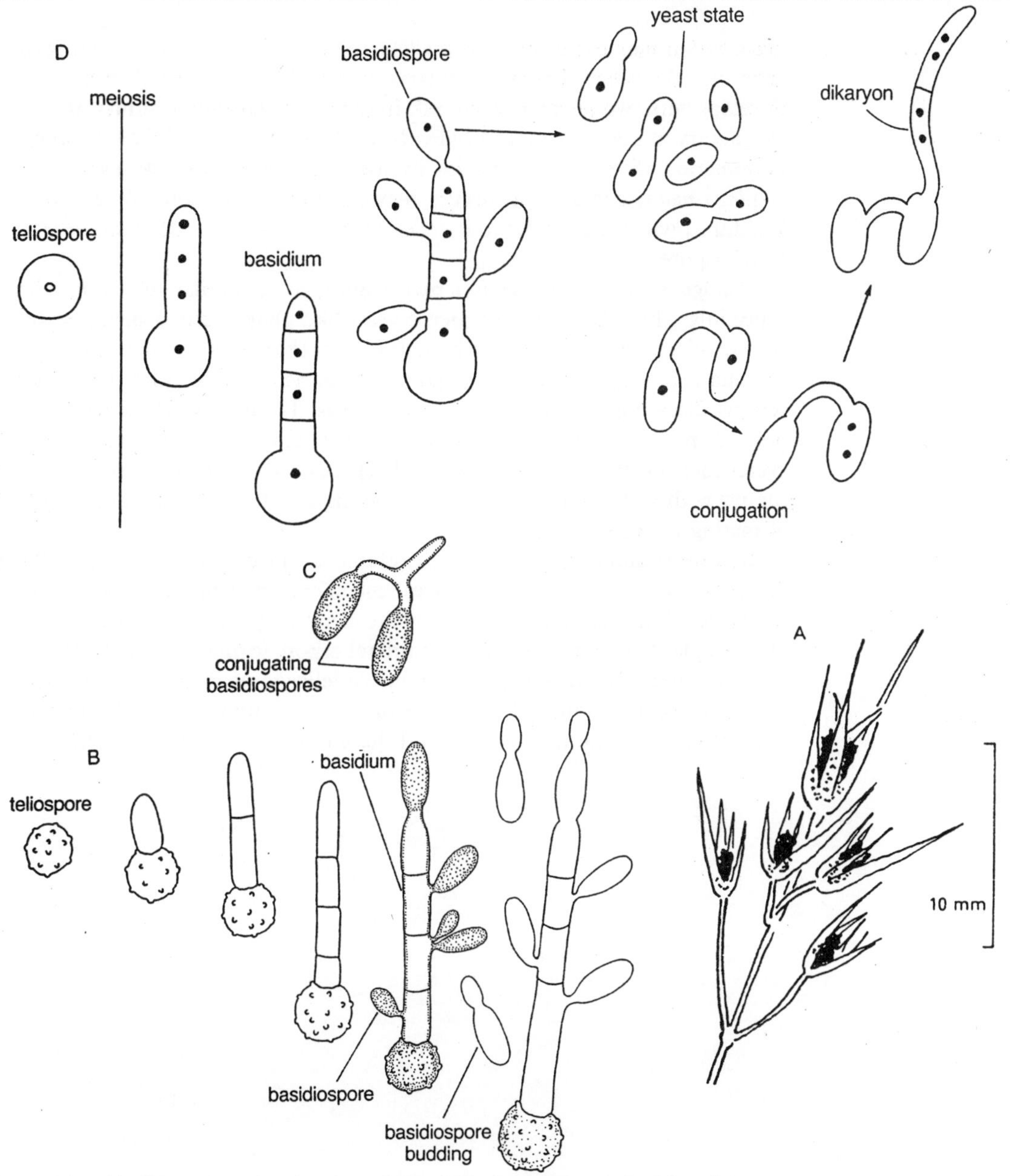

Fig. 5.26. *Ustilago avenae.* A, part of the inflorescence of *Arrhenatherum elatius* with the ovaries and associated tissue in each spikelet replaced by a sooty mass of teliospores. B, six stages in the germination of the teliospore over a period of 15 hours at 20°C; in the last stage two liberated basidiospores are budding and a second is forming from three of the cells of the basidium. C, liberated basidiospores conjugating. D, diagram of the process with nuclear details.

be maintained indefinitely in this state. If 'plus' and 'minus' basidiospores, or derived yeast-like cells of the two mating-types, are placed in contact, conjugation may occur, and the nucleus of one passes through a conjugation tube into the other. From the dikaryotic cell thus established, a mycelium develops with two nuclei in each cell. In the great majority of smuts, it is only this dikaryon that can infect a living host. The behaviour of the teliospore at germination differs from species to species. Sometimes, for example, in *Ustilago nuda*, conjugation occurs between 'plus' and 'minus' cells of the basidium, when it is at the four-celled stage, without the production of any basidiospores.

Ustilago maydis, which causes maize smut, has a similar life-cycle with a unicellular, haploid, non-pathogenic yeast-like phase, which develops from the basidiospores, a dikaryotic binucleate mycelial phase, which arises from this and is pathogenic, and a diploid phase restricted to the teliospores. It is a serious disease of maize in dry tropical areas. It forms galls on any above-ground parts of maize but especially the grain. These often become grotesquely enlarged and filled with teliospores. Currently it is a very popular fungus with molecular geneticists researching into the structure and function of mating type genes (Chapter 8).

Another common smut, to which reference will be made later (p. 128) is *Ustilago violacea* which parasitizes members of Caryophyllaceae (e.g. *Cerastium*, *Dianthus* and *Silene*). In white campion (*Silene alba*) for example, infected plants are not recognizable until teliospore formation which is limited to the anthers. These produce deep purple teliospores in place of the yellow pollen. Flowers of infected plants are most conspicuous because of the dark stain of scattered smut spores around the mouth of the white corolla.

Deuteromycotina 6

Deuteromycotina is an important division of fungi with around 17 000 species. However, it is a dustbin class. Into it are thrown species that have no known normal sexual stage so that they cannot be placed with confidence in other classes. Most reproduce by conidia, although a few are purely mycelial, developing no spores. The great majority of the fungi in the class are likely to be Ascomycotina in which the ascus stage has not yet been discovered or in which it has been lost in the course of evolution. A few may be conidial stages of Basidiomycotina. Theoretically when, as frequently happens, the ascus stage of a conidial fungus is discovered, it should be transferred to Ascomycotina. However, it is usually convenient to retain such a fungus in Deuteromycotina, if the asexual stage is the one usually to be found in nature.

In a fungus which has both an ascus (or basidium) stage and a conidial one, the former is known as the 'perfect' and the latter as the 'imperfect' state. This is why these fungi are commonly called Fungi Imperfecti. An added complication is that a particular fungus may have more than one imperfect state.

Using the more recent terms which were introduced into mycology the stage in which the normal sexual fusion of haploid nuclei occurs (e.g. ascus, basidium) is known as the 'teleomorph', the asexual stage being the 'anamorph'. The whole fungus including all its 'morphs' is known as the 'holomorph'.

In Deuteromycotina there is a great array of conidial apparatus. There are two main series: Coelomycetes and Hyphomycetes.

In Coelomycetes the conidia are often produced within a flask-like structure or pycnidium which is usually black and looks like a small perithecium. Members of this order are extremely common, particularly as saprotrophs on dead leaves and herbaceous stems. They appear as tiny black dots in the dead, bleached tissue. An example of this type, with rather large conidia, is illustrated in Fig. 6.1.

In Hyphomycetes the conidia are produced externally on conidiophores or, more rarely, directly from the vegetative mycelium. Two families are recognized: Moniliaceae in which the conidia and hyphae have walls that are unpigmented; and Dematiaceae in which some hyphal walls and especially those of the conidia are of a dark brown colour and tend to be thick.

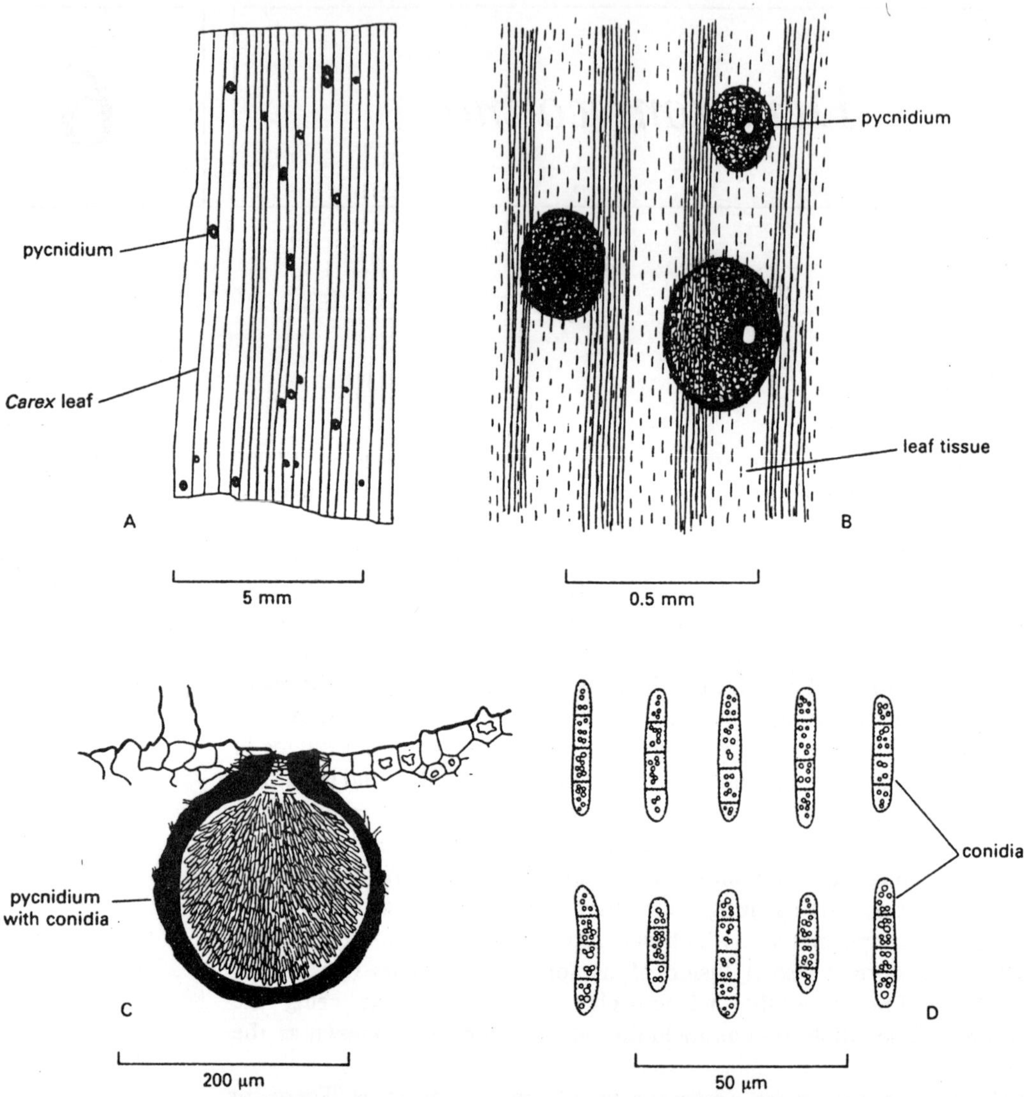

Fig. 6.1. *Stagonospora vitensis.* A, piece of dead *Carex* leaf with pycnidia. B, part with three pycnidia more highly magnified. C, section through a pycnidium full of conidia. D, mature conidia.

In these fungi there are several different patterns of conidial development. To illustrate this and to give some picture of Hyphomycetes a few examples, chosen for features of special interest, will be considered.

Botrytis cinerea (Fig. 6.2), grey mould, is extremely common as a saprotroph or as an unspecialized necrotrophic parasite. For example, it often attacks strawberries and raspberries, converting them to grey, mouldy masses. The conidiophore is a dark-walled, erect hypha, and the unicellular, oval conidia occur on lateral branches like bunches of grapes. At a certain level a group of bunches is formed, but the conidiophore continues to grow and at a higher

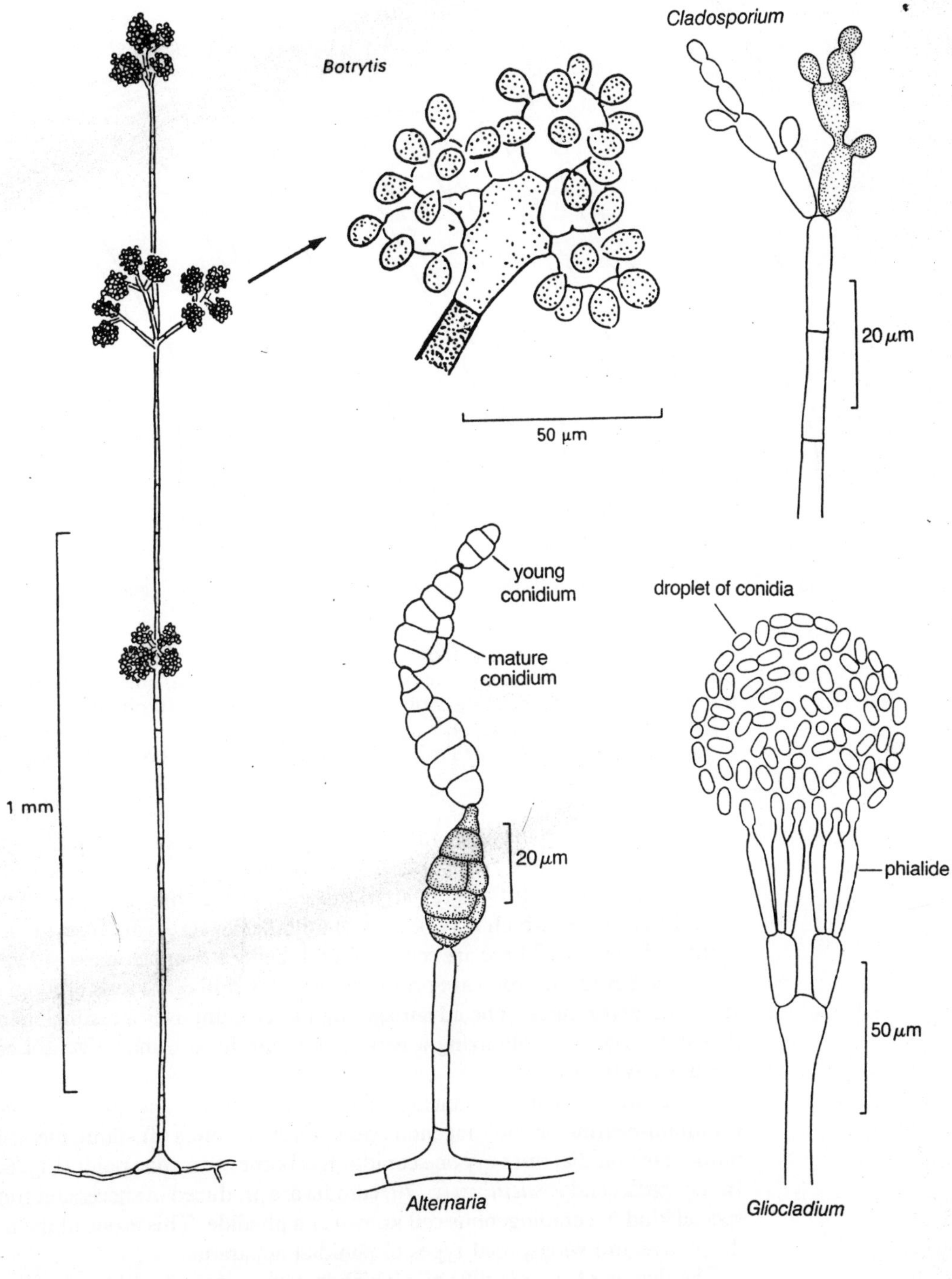

Fig. 6.2. Conidial apparatus of *Botrytis cinerea*, *Alternaria* sp., *Cladosporium herbarum* and *Gliocladium roseum*.

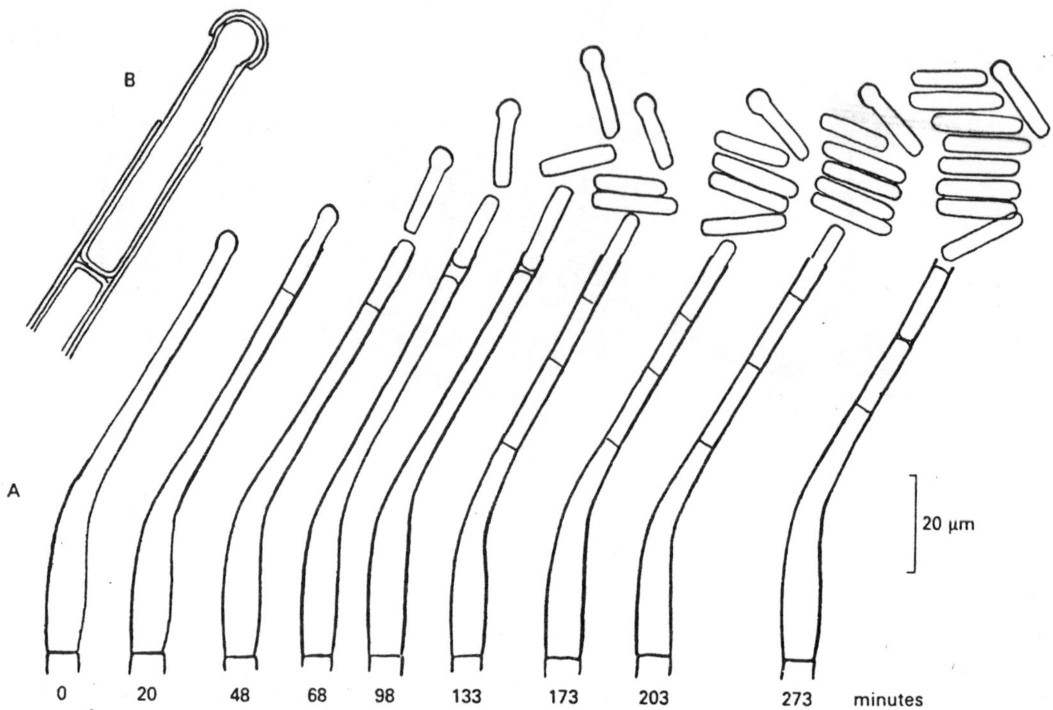

Fig. 6.3. *Thielaviopsis basicola.* A, phialide drawn at indicated times over a period of 273 minutes. At the start the outer wall is intact, but 20 minutes later the outer wall has ruptured near the tip and the first phialoconidium is emerging carrying a cap of outer wall, details of this are shown at higher magnification in B. After 273 minutes nine phialoconidia have been liberated.

level another group may be produced. The conidia are dry and easily blown away. On nutrient agar, the conidial stage is developed in abundance and the whole culture has a fluffy, grey appearance. Later, as this growth collapses, hard, black bodies which may be several millimetres across are formed on the surface of the agar. These are sclerotia, each being a compact mass of hyphae with a hard rind. Sclerotia are produced by many different kinds of fungi and are resting structures. The advantages of a sclerotium over a resting-spore is that it is larger. In colonizing a new substratum its size may give it a competitive advantage.

In *Botrytis* the lateral branches of the conidiophore end in one or more conidium-bearing or conidiogenous cells. Each cell buds off simultaneously a number of conidia, but only one conidium is borne at any one point (Fig. 6.2). In *Aspergillus* and *Penicillium* (p. 70) conidia are produced in succession from a special kind of conidiogenous cell known as a phialide. This is one of the most distinctive and widespread types of conidial apparatus.

The details of the activity of a phialide can be described by reference to *Thielaviopsis basicola* (Fig. 6.3) which causes a root-rot of tobacco. The phialide, formed at the end of a short lateral branch of the mycelium, is in the

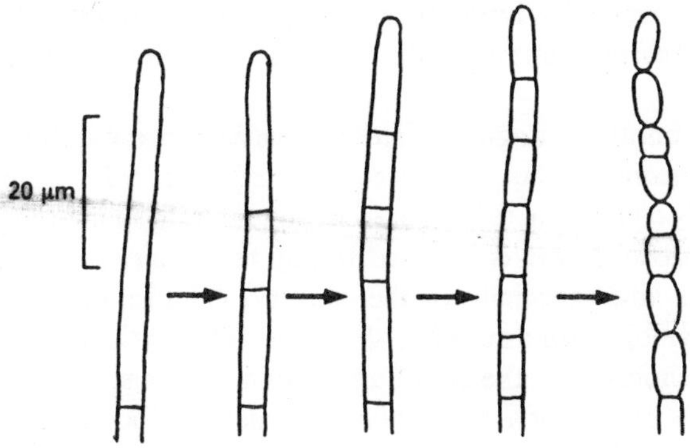

Fig. 6.4. *Geotrichum candidum*. Time-lapse sequence of arthroconidia formation in a terminal hypha. (After photographs by Cole and Kendrick)

form of a narrow flask with a long neck. When the first conidium is about to be produced, the outer layer of the cell wall ruptures near the tip of the neck and the conidium, surrounded by a wall derived from the inner layer of the phialide wall, is pushed through the open tubular beak of the outer wall. The first-formed conidium bears a capping of outer wall, but this is lacking in those that follow. The conidia are delimited at some depth within the phialide, but in most phialides (e.g. in *Aspergillus*) the cross-wall marking the base of the conidium is formed at the mouth of the phialide, which has a slightly thickened rim.

In *Aspergillus* and *Penicillium* a phialide gives rise to a long chain of phialoconidia with the youngest at the base. However, in some Deuteromycotina slime is produced with the phialoconidia which, instead of producing dry chains, form slimy masses. Biologically the difference between dry and slimy conidia may be of considerable significance. The former are easily dispersed by wind; the latter cannot directly be blown away and depend on insects or rain splash for dispersal. *Gliocladium roseum* (Fig. 6.2) is an example of a mould with a slimy conidia. Basically the conidial apparatus resembles that of *Penicillium*. However, the conidia produced by the phialides, instead of forming dry chains, run together into a single drop of slime.

In being surrounded by a wall which is derived from the inner layer only of the wall of the conidiogenous cell, phialoconidia differ from other kinds of conidia. In these other types the conidial wall originates from both layers of the parent cell wall.

Some Hyphomycetes form chains that are not produced by phialides. Thus in *Geotrichum candidum* a chain is formed by the fragmentation of a hypha following the development of a number of cross-walls (Fig. 6.4). Such conidia are known as arthroconidia.

Another type of chain formation is seen in *Alternaria* (Fig. 6.2), a large genus of moulds found in abundance on decaying vegetation and on other damp cellulosic materials. The multicellular conidium is formed at the end of a hypha, and then another is produced by budding from the apex of the first. Repetition of this process leads to a short chain of conidia with the oldest at the base. Conidia formed by budding are termed blastoconidia. The conidial chains of *Alternaria* contrast with clusters of solitary blastoconidia of *B. cinerea*.

A still more abundant fungus is *Cladosporium herbarum* which forms similar chains of blastoconidia, but they are branched (Fig. 6.2). Conidia of this species normally contribute the major element of the air-spora (see p. 125). If a petri dish of nutrient agar is exposed briefly to the air, particularly in the warmer months and especially in the country, colonies of a velvety green mould soon develop, outnumbering all others. This is *C. herbarum*.

In many Hyphomycetes the conidium is solitary on a simple conidiophore or at the end of a branch of a more complex structure. *Clavariopsis aquatica* is an example (Fig. 6.5). Here the conidium starts as a swelling at the end of a conidiophore and is soon delimited by a cross-wall. When fully grown it leaves the conidiophore by rounding-off at the wall separating the two. The vacated conidiophore does not immediately give rise to another conidium, as happens in a phialide, and both the outer and inner layers of the conidiophore wall contribute to the cell-wall of the conidium. Such conidia are termed aleurioconidia.

Although most Hyphomycetes are terrestrial, many being parasites of seed plants while many more occur as saprotrophs on all sorts of dead organic matter, a number grow completely submerged in water. An interesting assemblage of these fungi is found in abundance and with great regularity on submerged, decaying leaves of deciduous trees (e.g. alder, willow, oak) especially in well-aerated streams and rivers. They are particularly abundant in autumn and winter. The mycelium ramifies in the dead leaf tissue and the conidiophores grow out into the water. The production, liberation and dispersal of the conidia normally occurs below the surface. These fungi seem to be the principal agents of decay of these leaves which represent the major annual input of organic matter in a stream. Evidently they play a vital part in stream ecology.

A special feature is that in most species the conidium is branched, and often it takes the form of four long arms diverging from a common point. Some of these are phialoconidia, others are not (Fig. 6.5). However, some aquatic Hyphomycetes have unbranched spores which are long and worm-like with curvature in more than one plane.

What appears to have been the repeated evolution of the tetraradiate spore, indicated by different patterns of development, suggests a biological advantage. Indeed, there is a strong probability that such a conidium has survival value as a miniature anchor. Arrest on a suitable substratum is a real problem for spore dispersal in a stream. Like branched conidia, those that are long and worm-like have a better chance to come to rest on a submerged leaf in running water than those that are spherical or ovoid.

In a well-aerated stream, conidia of aquatic Hyphomycetes reach enormous numbers in autumn, up to 20 000 per litre of water. They tend to be trapped by foam. If persistent foam from a babbling brook is collected, fixed (to stop germination) and examined microscopically, these conidia are easily seen. A

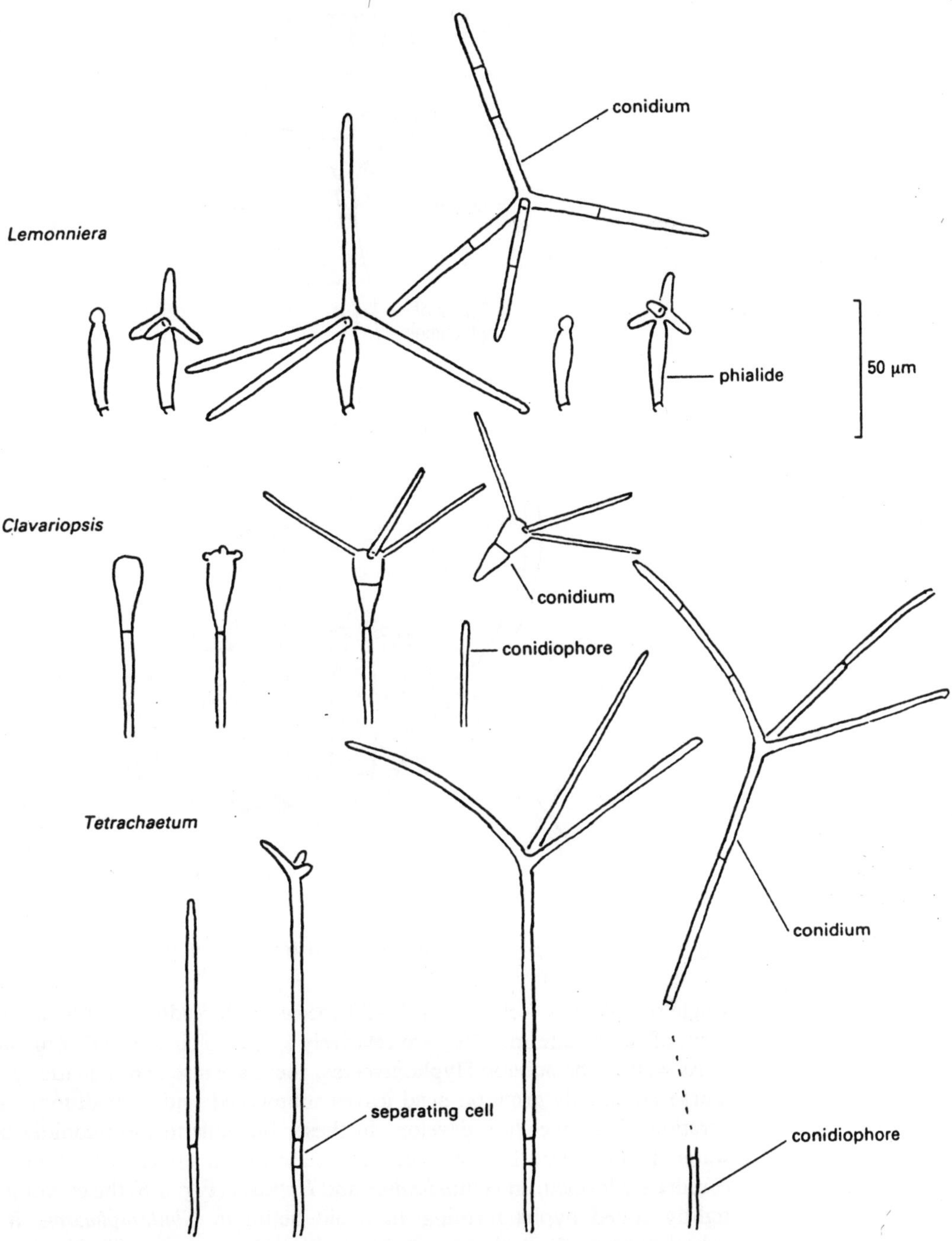

Fig. 6.5. Aquatic Hyphomycetes. *Lemonniera aquatica*: stages in phialoconidium formation from a phialide. *Clavariopsis aquatica*: stages in conidium development, only one conidium is formed and liberated from the conidiophore by separation at a cross-wall. *Tetrachaetum elegans*: only one conidium is formed and this is liberated by the breakdown of a separating cell.

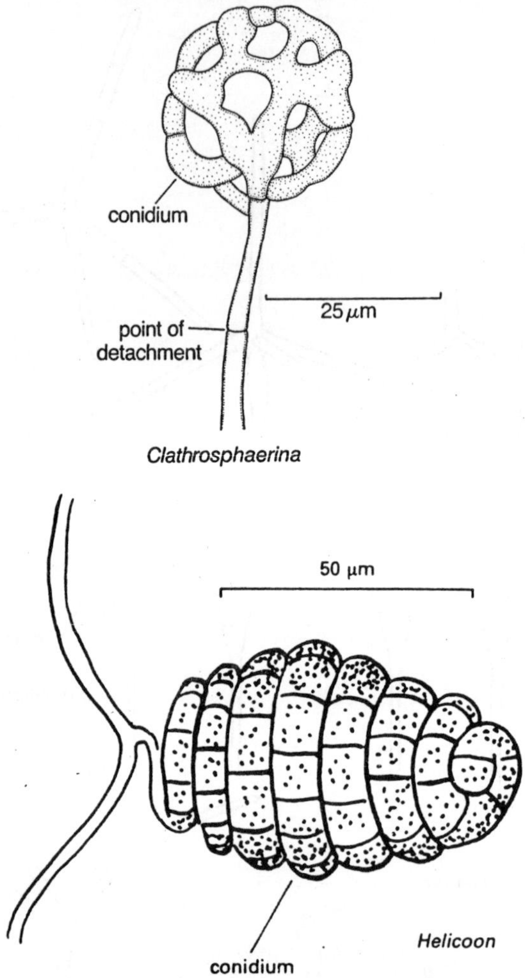

Fig. 6.6. *Clathrosphaerina zalewskii* and *Helicoon richonis*.

single sample may well contain 10–20 species each readily identifiable by the form of the conidium. They are relatively large, colourless and thin-walled.

As well as the aquatic Hyphomycetes, there are also aero-aquatic species. These commonly grow on dead leaves submerged under conditions of low aeration. The mycelium develops in these, but fails to form conidia below water. If, however, the leaves are exposed to the air under moist conditions, conidia are formed. In *Helicodendron* and *Helicoon* (Fig. 6.6) the conidium is a tightly coiled hypha forming an ovoid helix; in *Clathrosphaerina* it is a spherical network. Both types trap air within the structure. The fact that the conidia cannot be wetted precludes dispersal below water, but they would seem well adapted to spread by floating on the surface.

Certain conidial fungi are of special interest because they capture and consume eelworms. They appear to be common in soil, leaf litter, rotten wood and the dung of herbivores. One of the commonest species is *Arthrobotrys*

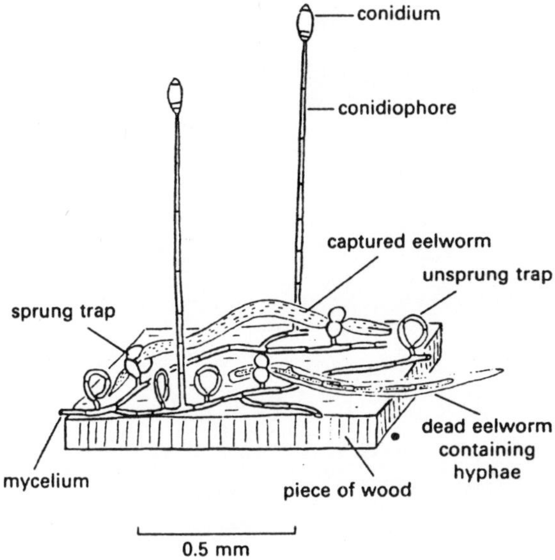

Fig. 6.7. *Dactyella bembicodes.* Diagrammatic sketch of fungus growing on a piece of damp wood. Seven ring-traps are shown, three of them sprung. One nematode, caught near its head, has hyphae inside it. A second, caught head and tail, has not yet been invaded. Two conidiophores are shown, each with a single conidium. (Modified after Couch)

oligospora. This has a branched, septate mycelium small portions of which may be organized as three-dimensional networks. In these the outer walls of the hyphae are coated with a strongly adhesive substance. An eelworm nosing through one of these labyrinths gets irretrievably stuck and, after fruitless struggles, dies. Hyphae then penetrate the eelworm and consume its substance. Just how far *Arthrobotrys* in nature is dependent on nematodes is not clear. It can grow and sporulate freely on ordinary laboratory media, traps being normally produced only when eelworms are present. There is considerable variety in the trapping devices of these predaceous fungi. The most spectacular is found in *Dactylella bembicodes* (Fig. 6.7) and in a number of other species. Here the trap is a short-stalked, three-celled ring like a noose-snare for a rabbit on a microscopic scale. When an inquisitive eelworm enters this ring, it stimulates the three cells which suddenly inflate gripping the animal. The inflation of the cells is completed in a tenth of a second and involves a three-fold increase in their volume. How this dramatic increase is achieved has not yet been fully explained.

Dispersal in fungi 7

Dispersal is important for a fungus, as it is for any organism, in order to maintain the species in its existing range and perhaps to extend that range, and also to spread genetic variability, as it arises, throughout the population. In fungi the feeding mycelium is usually concealed in the nutrient substratum and what we normally see are the structures concerned with the production and liberation of spores which are the dispersive units. In most fungi spores are wind-borne. In aerial dispersal, as with aircraft, three episodes can usually be recognized: spore release (take-off); actual dispersal (flight); and deposition (landing).

In earlier chapters there has been frequent reference to spore release. In many fungi this is an active process, the spores being shot into the air. This is the rule in most Ascomycotina and in Hymenomycetes, the largest class of Basidiomycotina. In Hymenomycetes the spore-gun is a basidium with a range of only 0.1–0.2 mm, but in Ascomycotina the range of the ascus is much greater varying, according to the species, from 2–300 mm.

In both Ascomycotina and Basidiomycotina the first step in effective dispersal is the launching of the spores into air that is likely to be turbulent. In contact with the ground is a thin layer of air that is still or in laminar flow and any spores set free into this layer settle rapidly. Commonly it is a few millimetres deep, although, particularly during calm nights, it may thicken considerably, while under the highly turbulent conditions of midday, it may be reduced to a fraction of a millimetre. Above this layer air is turbulent and spores reaching this region stand a good chance of wide dispersal.

Most asci shoot their spores to a sufficient distance to reach the potentially turbulent air, but the range of the basidium is too short for this to occur. However, with the pileus of a toadstool on top of a stipe, or with a bracket fungus on a tree, the spores, shot from hymenia that usually are nearly vertical, can then drop freely into air that is often turbulent. This situation is illustrated in Fig. 7.1 which contrasts spore liberation from an agaric with that from a cup-fungus belonging to Ascomycotina.

One other difference between an agaric and a cup-fungus should be noted. The layer of asci and paraphyses in the latter is little affected by rain, but the hymenia of a toadstool are ruined if directly wetted. Thus the shelter that the

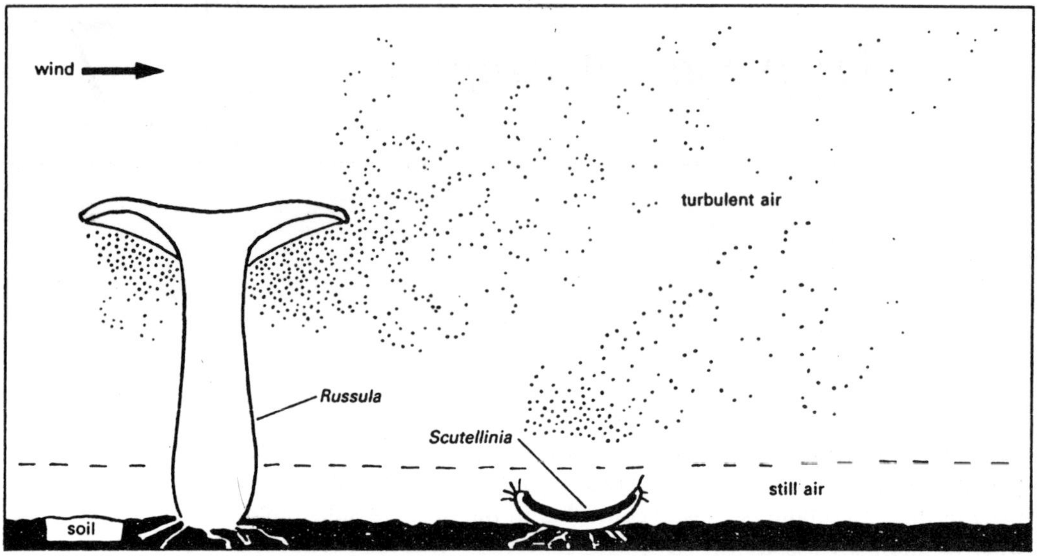

Fig. 7.1. Diagram of the downwind drift of airborne spores from an agaric (*Russula*) and from a cup-fungus (*Scutellinia*) which has just puffed. The air in contact with the soil is still; above it is turbulent.

cap of a toadstool provides for the delicate hymenial surfaces is significant. The umbrella-like form is no accident but has real survival value.

In Hymenomycetes and in most Ascomycotina, spore discharge, depending as it does on the activity of turgid cells, requires a sustained supply of water. Many Ascomycotina together with leathery and gelatinous Basidiomycotina, so common on wood and bark, cease to liberate their spores on drying, but remain alive. They quickly absorb water when wetted and discharge begins again. The perithecial stage of *Nectria* (p. 65), the brackets of *Stereum* (p. 77) and the gelatinous basidiocarps of *Auricularia auricula-judae*, abundant on dead elder (*Sambucus*), are common examples.

Although violent spore discharge in fungi is mostly dependent on damp conditions, there are a few in which the discharge is associated with drying. In certain species of *Peronospora* (p. 45) the liberation of sporangia in the field is known to be associated with rapid reduction in the relative humidity of the air. Discharge of sporangia was originally attributed to the twirling on drying of the main axis of the sporangiophore, thus throwing off the finely attached sporangia. However, evidence has recently been presented suggesting that discharge is an electrostatic effect. During rapid drying leaves of the host plant become charged and apparently repel the similarly charged sporangia. There is no doubt about the leaves becoming charged, nor that the liberated sporangia also carry a charge. There is, nevertheless, difficulty in accounting for the initial take-off. Once airborne, however, the charged sporangia are certainly repelled.

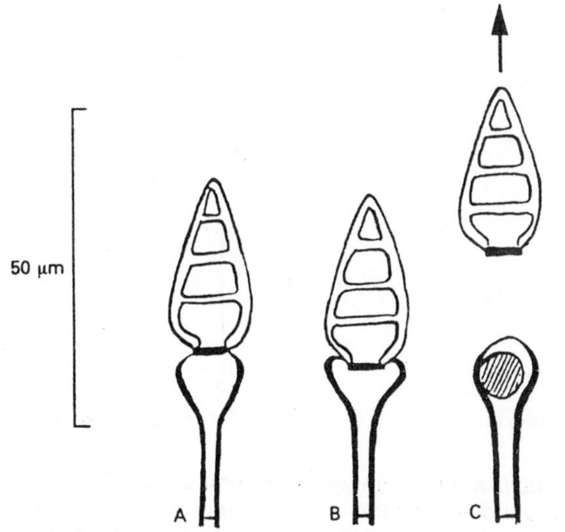

Fig. 7.2. *Deightoniella torulosa.* A–C, stages in drying leading to spore discharge; a fraction of a second separates stages B and C. The gas phase in the terminal cell of the conidiophore in C is shaded. (Modified after Meredith)

In a few conidial fungi a discharge mechanism occurs which depends on the rupture of water under stress. For, example, in *Deightoniella* (Fig. 7.2) a single conidium occurs on a conidiophore the top cell of which is unequally thickened. Its apical region is thin-walled, but elsewhere the wall is much thicker. On exposure to dry air, water evaporates from this cell and, to accommodate the resulting decrease in volume, the apex is sucked inwards. However, the elastic wall is straining to return to its former position. As a result the aqueous contents of the cell are under increasing tension. Eventually the strain becomes too great and the cohesion of the water molecules, or their adhesion to the wall, is overcome. In a flash, as a bubble of gas appears, the cell returns to its former shape, the water having ruptured. This sudden movement jerks the conidium from its conidiophore. This type of discharge has been reported in a small number of other conidial fungi. The same mechanism of water rupture is responsible for the violent discharge of spores from the sporangium of a fern.

In many fungi there is no active discharge; nevertheless, the spores are freely liberated into the air. Well-known examples amongst moulds are *Cladosporium*, *Penicillium*, *Botrytis* and *Rhizopus*. Again theurediniospores of rusts (Uredinales), the teliospores of smuts (Ustilaginales) and the conidia of powdery mildews (Erysiphales) are often abundant in the air, although there is no violent spore discharge. However, in moulds the spores are usually produced several hundred micrometres above the substratum on erect conidiophores or sporangiophores and this, no doubt, aids take-off. Further the spores of rusts, smuts and powdery mildews are mostly borne on diseased

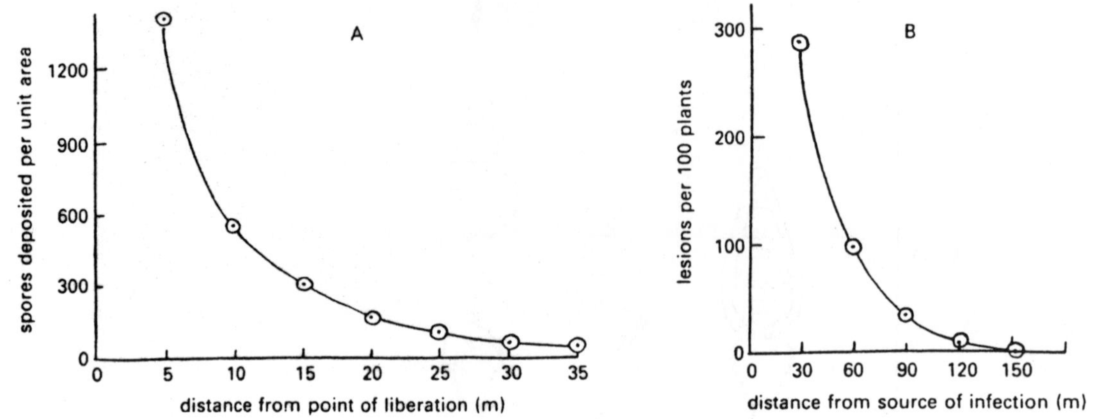

Fig. 7.3. A, *Podaxis* spores deposited per unit area (147 mm^2) on horizontal slides on the ground downwind at increasing distances following a single massive liberation of spores at the zero position and 0.5 m above ground level. (Data from Sreeramulu and Ramalingan 1961.) B, *Phytophthora infestans*, primary blight infections (recorded as lesions per hundred plants) in a potato field near an infected potato-tuber refuse pile providing the source of infection. (Data from Bonde and Schulz 1943)

plant surfaces well above ground level, so that spores are normally liberated into air that tends to be turbulent.

Most air-borne spores are small, often about 10 μm in diameter, and their rate of fall in still air is less than 10 μm s^{-1}, more than ten times slower than that of thistle down. Consequently their passage in normal turbulent air is determined largely by the mass movement of the air and only to a slight extent by gravity.

After take-off spores are spread by the wind. The horizontal smoke-plume from a tall factory chimney provides a model aerial dispersal of minute particles from a point source. The smoke moving downwind is diluted by mixing with peripheral smoke-free eddies of the turbulent air and thus the plume assumes a form that is often more or less conical. A spore cloud, though too thin to be visible, may be envisaged as somewhat similar, but it is normally liberated at a lower level so that on drifting downwind its base drags along the ground depositing some of its load of spores. The number of spores deposited on unit area declines steeply from the source. A typical situation is illustrated in Fig. 7.3A. This curve of dispersal is important for a plant pathologist attempting to define the practical limits of a danger zone around a source of fungal infection in a crop. The actual infection curve has a regular form (Fig. 7.3B) no doubt reflecting the rapid decline of spore deposition on receding from the source of inoculum.

As well as studying the horizontal dispersal of spores from a point of liberation, mycologists have investigated vertical distribution. To this end some workers have trapped spores on sticky slides exposed from aircraft at varying heights above wheatlands heavily rusted with *Puccinia graminis*. For example, on one occasion in August over Canadian wheatfields, 3870 urediniospores were trapped per square centimetre of slide surface in 10 minutes at a height of 300 m, 1210 at 1500 m, 17 at 3000 m and 2 at 4200 m.

It seems that rust spores are carried upwards by thermal turbulence and form veritable spore clouds which may drift for hundreds of kilometres causing infection far from the region of origin. There is good evidence that the early summer infection of the wheat crop in Canada is often due to urediniospores of *Puccinia graminis* carried northwards from zones 500–10 000 km to the south where rusting occurs earlier in the year.

An important consideration in dispersal is the viability of the spores. There is biologically no significance in distant dispersal if at the end of the journey the spores are no longer alive. During aerial dispersal they are subject to the possible injurious effects of desiccation and strong light, especially ultraviolet; and their ability to withstand these adverse conditions differs greatly. For example, the urediniospores of rusts remain viable for periods of many days during aerial transport, while the basidiospores soon lose their ability to germinate. Thus these spores normally infect only nearby hosts while urediniospores can carry infection over great distances.

A striking feature of most fungi is the enormous spore production. A big specimen of the giant puff-ball (*Calvatia gigantea*) may produce seven million million spores, and medium-sized mushroom (*Agaricus campestris*) may liberate 500 000 a minute throughout a spore-fall period lasting several days. The wastage is clearly enormous. The factors that impose this wastage may be: the infrequency of suitable niches existing under the right conditions for colonization; the rapid decline in spore deposition on receding from a point of liberation; loss of viability during dispersal; and the frequent need for two spores of compatible mating-type to germinate together if a sporulating mycelium is to result. It is almost impossible to give quantitative values to these factors, but the net result is that the chances against an individual spore succeeding in its reproductive function may be astronomically high.

Fungi are so abundant, and the production of their spores is on such a large scale, that these form a normal constituent of the air. Much attention has been given to the 'air-spora', defined as the population of pollen grains and spores, mainly fungal, suspended in the air; and this study is known as 'aerobiology'. In investigating the air-spora most aerobiologists now use some form of volumetric spore trap which allows concentrations of spores to be estimated, usually expressed as the number in a cubic metre of air.

So far as the fungal element of the air-spora is concerned, this is of special interest in plant pathology. If the spores of an important pathogen can be recognized and their numbers estimated, local warning to spray with fungicides can be properly timed. Pollen grains and fungal spores are also of importance in inhalent allergy and much of the study of the air spora has had a medical incentive.

Most aerobiologists use some form of impactor trap in which the spores, impacted on a sticky surface, can then be identified under the microscope. The Hirst spore-trap (Fig. 7.4) has been widely used. Air is drawn by a suction pump at a measured rate (10–20 litres per minute) through a horizontal slit in an otherwise air-tight system, and the slit is kept facing into the wind by a

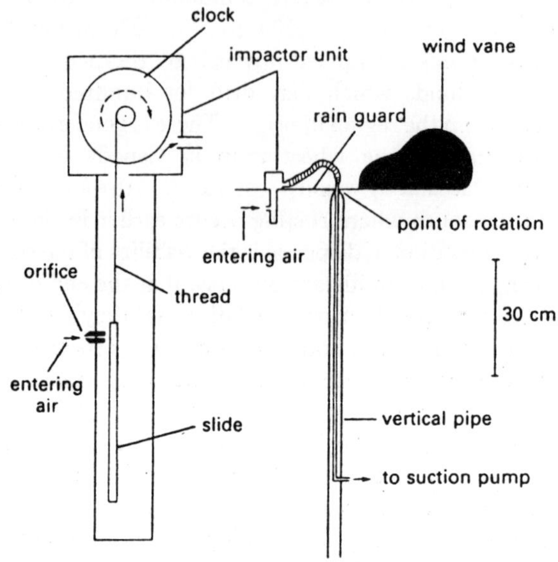

Fig. 7.4. Hirst spore-trap. Right: set-up as seen in vertical section. Left: very diagrammatic enlarged representation of impactor unit to show mode of operation.

large vane. Within the system spores are impacted on a sticky microscope slide drawn upwards by clockwork past the slit at 2 mm per hour. The form of a slit and its exact relationship to the slide are critical for the efficiency of the trap. The slide is replaced daily. From spore counts on a defined area of the slide, and from a knowledge of the volume of air that has impinged on that area, the concentrations of each kind of recognizable spore in the air can be calculated throughout the 24-hour period. Figure 7.5 gives some idea of the appearance of a small area of a slide from a Hirst trap.

A number of variants of this automatic volumetric spore trap, have been introduced. One, the Burkard spore trap, works on exactly the same principle but has a rotating trace of adhesive-coated transparent plastic on a drum which rotates once in 7 days. Such traps are about 80% efficient down to spore sizes of 3 μm and are used where total atmosphere spore counts are required.

Impactors are also used for culture studies where viable spore counts are required. One of the best known of these is the Andersen sampler (Fig. 7.6). Air is drawn through a circular orifice and then through a stack of six circular plates, each perforated by 400 holes. Successive plates down the stack have progressively smaller holes. A petri dish of nutrient agar is placed below each plate. As the spores are pulled through they become separated into six aerodynamic sizes and impact onto the nutrient agar. The largest are deposited on the first dish and the smallest on the last. The spore trap is obviously only useful for spores of fungi which will grow in culture and, where atmospheric spore concentrations of these are dense only a very limited volume of air can be sampled, otherwise the colonies formed from the

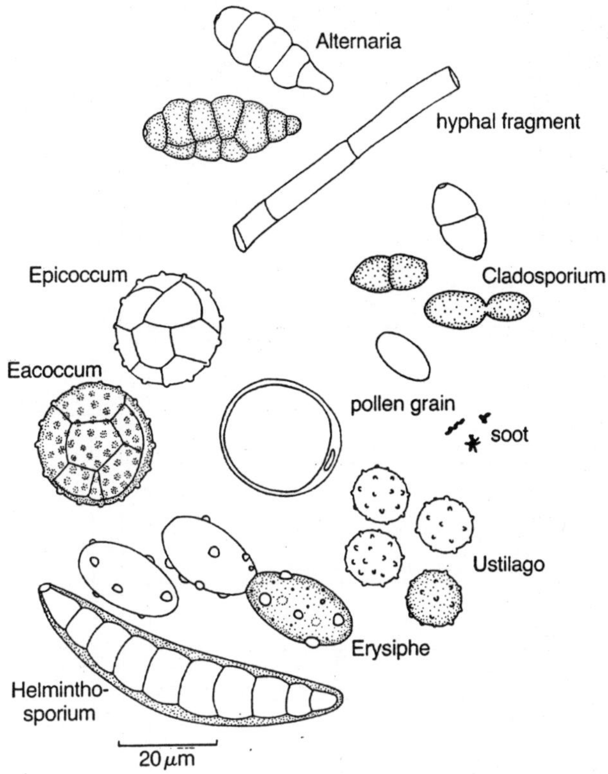

Fig. 7.5. Elements of London air-spora in June as seen in an H.P. field on examining a slide from a Hirst spore-trap under the microscope. There are pollen grains, various fungal spores, hyphal fragments and numerous irregular particles of soot.

impacted spores overgrow and outcompete each other. In practice it has been most successfully used in allergy studies involving bacteria, Actinomycetes and fungi in mouldy hay and for studying minority components of the air-spora which have specific nutrient or temperature requirements, such as osmophilic aspergilli and thermophilic fungi. In these instances use of high sugar nutrient media or high incubation temperatures eliminate the abundant common moulds.

In most parts of the world by far the commonest spores in the air are those of *Cladosporium*, a mould abundant on decaying vegetation. Spores of other moulds (e.g. *Alternaria, Epicoccum*) are also present. In country areas in summer the conidia of powdery mildews and the urediniospores of rusts abound in the air-spora, the nature of which changes throughout the day (Fig. 7.7). In particular it tends to be dominated around midnight and in the early hours by ballistospores of mirror-image yeasts, particularly *Sporobolomyces*. Conidia of *Cladosporium*, sporangia of *Phytophthora infestans* and the teliospores of *Ustilago* have their maximum in the midday hours (Fig. 7.7).

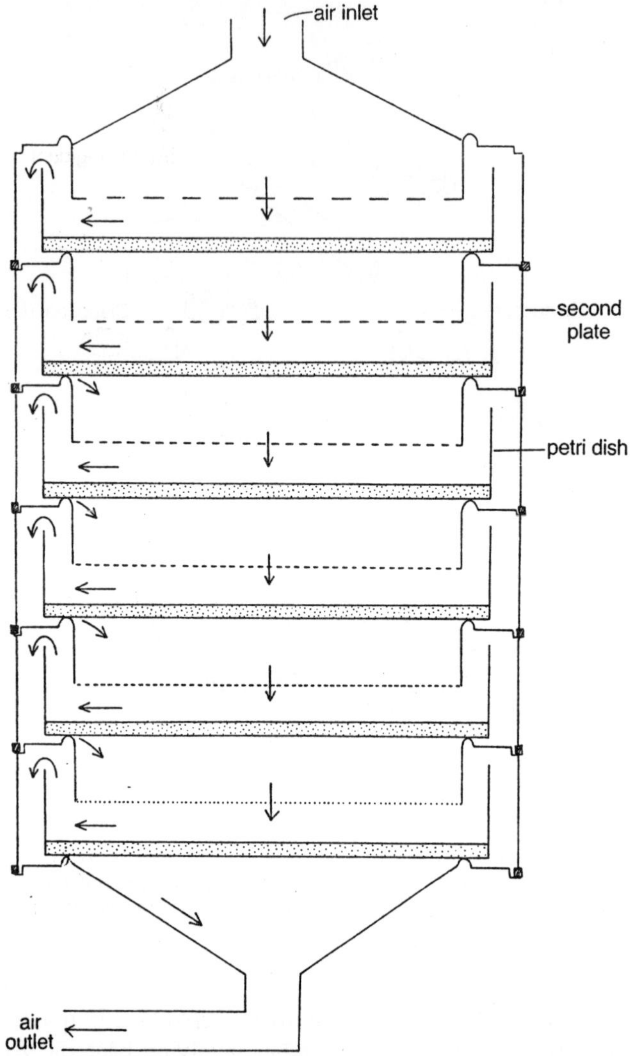

Fig. 7.6. Andersen sampler. Simplified sectional view. Airflow through sampler indicated by arrows.

Sporangia of *Peronospora* peak shortly after dawn. These patterns reflect, in general, the times when the spores are liberated, but may be apparent only when the situation is averaged over a number of days.

The air-spora is much influenced by the weather and in particular a dry air-spora and a wet one have been recognized. In dry weather it is rich in spores of moulds, in urediniospores of rusts and in the conidia of downy and powdery mildews. Heavy rain has the immediate effect of causing leaf flutter and stem vibration which shakes spores from diseased plants and from dead mouldy vegetation, thus enriching the dry air-spora. However, such heavy rain soon scrubs spores out of the air and deposits them on the ground. Thereafter a wet air-spora develops dominated by elongated ascospores discharged from the

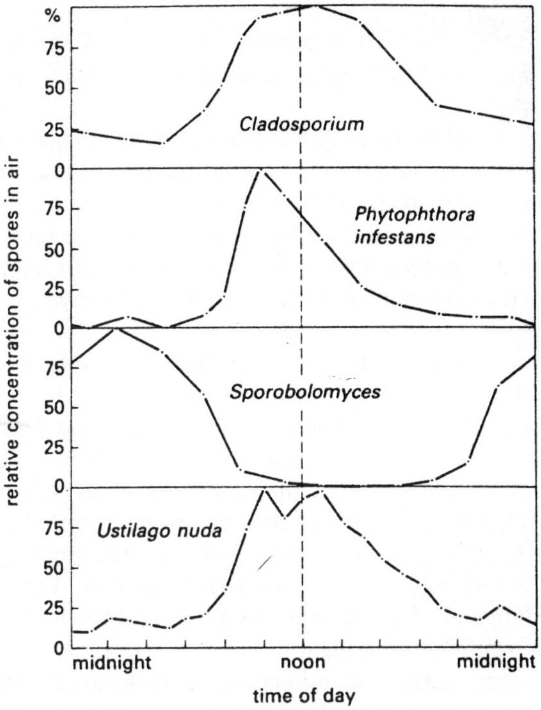

Fig. 7.7. Spore content (Hirst-trap determinations) throughout the day as percentage of the geometric mean concentration. Curves for *Cladosporium* and *Phytophthora* based on observations on agricultural land at Rothamsted (Hirst 1953), for *Sporobolomyces* curve based on work at Thornley Island, Chichester Harbour (Gregory and Sreeramulu 1958), *Ustilago* curve based on observations in infected barley field in Berkshire (Sreeramulu 1962)

many species of Ascomycotina that grow on dead vegetation and in which spore discharge depends on wetting. The wet air-spora is normally of brief duration.

The air-spora varies throughout the year, many more spores being present in the summer than during the winter months. Again in country areas it is richer than in towns.

Deposition is the last stage in the story of aerial dispersal. Spores may be deposited by the direct action of gravity, but probably more often by impaction under the influence of eddies in the air. If a microscope slide, made sticky on both surfaces, is exposed horizontally a short distance above ground level, it is often found, on microscope examination, that nearly as many spores occur on the under as on the upper surface. However, when non-turbulent conditions obtain, deposition is only on the upper surface.

In connection with impaction in turbulent air, the size of the spores is important. Small spores (5 μm or less in diameter) in an air-stream tend to flow past an obstacle, while larger ones (10 μm or more in diameter) are much more readily impacted. It has been remarked that many pathogens of crop plants (especially rusts, powdery mildews and downy mildews) have relatively large spores that are good impactors on stems and leaves. In contrast such soil fungi as *Aspergillus* and *Penicillium* have small spores with a low impaction efficiency.

There are some fungi in which the spores occur in an aqueous slime and cannot be set free directly into the air by wind. A few of these are spread by insects, but for many rain-splash is involved. Species belonging to such genera as *Fusarium*, *Colletotrichum* and *Gliocladium* are common examples, and also the conidial stage of the coral-spot (Fig. 4.13). These are sometimes called slime-spore fungi in contrast to dry-spore types such as *Botrytis*, *Cladosporium* and *Penicillium*.

The basic features of splash dispersal have been studied by allowing large drops of water to fall on to films of water containing spores, and analysing the reflected splash which contains liquid both of the impinging drop and of the target film. It was found that a drop 5 mm in diameter falling from a height of 7.4 m on to a thin film of water containing abundant conidia of *Fusarium solani* produced over 5000 reflected droplets varying in diameter from 5 to 2400 μm. They were thrown to a distance of up to 1 m and a large proportion of them carried conidia. Splash dispersal is a short-range process, but some of the smallest spore-containing droplets may remain suspended in the air and be dispersed by the wind. In some plant diseases, where slime-spore fungi are concerned, the combination of rain-splash and wind is important in epidemiology. A specialized example of splash dispersal is to be seen in the bird's nest fungi belonging to the two common genera *Cyathus* and *Crucibulum* (p. 97).

Although relatively few in number, there are many examples, widely scattered taxonomically, of fungi dispersed by insects of all types: flies, beetles, moths, wasps.

The basidiospores of *Phallus*, the pycniospores of rusts, and the conidia of ergot rye (*Claviceps purpurea*) are all minute slime-spores produced in an aromatic, sugary matrix and normally spread by flies.

A particularly interesting example of insect dispersal is *Ustilago violacea*. This fungus attacks species of *Silene*. In white campion (*S. alba*) there are separate staminate and pistillate plants. The flower of the latter has an ovary but only staminal rudiments. When, however, a pistillate plant is infected by the smut, these rudiments develop into stamens of a fairly normal form, but their anthers dehisce to liberate minute purple–black smut spores instead of the relatively large, yellow pollen grains (Fig. 7.8). As in other smuts, although a considerable part of the shoot is infected with no outward indication of the fungus, spore formation is sharply localized. The infected flowers, commonly to be seen in summer, are conspicuous on account of the purple–black anthers. It seems that the spores are spread to healthy plants by night-flying moths.

Another example is the spread by bark-beetles of *Ceratocystis ulmi* causing Dutch elm disease. The beetles form characteristic breeding galleries at the interface of bark and wood in recently killed elm trunks, and on the walls of these galleries the fungus sporulates. Beetles of the new generation emerge through bore-holes in the bark already contaminated with spores and, in

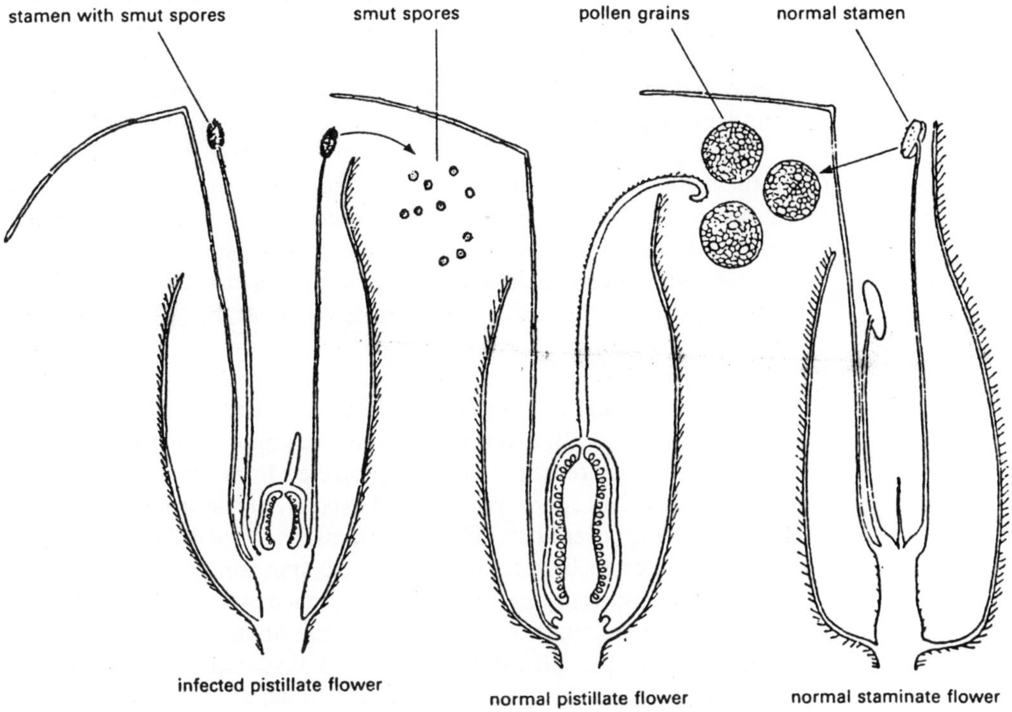

Fig. 7.8. Flowers of *Silene alba* in longitudinal section. Middle, pistillate flower with minute staminal rudiments just below the ovary. Right, staminate flower. Left, flower from a pistillate plant infected with *Ustilago violacea* in which the staminal rudiments have developed into stamens but with smut spores in the anthers. Smut spores and pollen grains highly magnified to the same scale.

feeding on twigs of living elms, introduce the lethal pathogen to its host. Later the beetles, still contaminated, bore through the bark of dead trunks and inoculate the new breeding galleries with the fungus.

A more casual example of insect dispersal is to be found in orchards in autumn. *Monilina fructicola* is responsible for the brown soft-rot of apples. Wasps pick up spores while crawling over rotting fruit covered with spore pustules. Flying to healthy apples they bite through the protective cuticle and introduce spores to the soft tissue which is rapidly invaded by the mycelium. The spores of *M. fructicola* are not, however, spread only by wasps for they are dry and powdery and can be wind-dispersed.

Some fungi rely on mammals for dispersal. This is particularly true of the hypogeous fungi of which truffle (*Tuber*) and Hart's truffle (*Elaphomyces*) are well-known examples. These are Ascomycotina. However, underground reproductive structures (basidiocarps) have also been developed in certain Basidiomycotina (e.g. *Hymenogaster* in Gasteromycetes) and in some Mucorales (e.g. *Endogone*). All these when ripe give out a smell which attracts small mammals which grub them up to eat. In woods in Britain it is common to see shallow holes where squirrels have dug for *Elaphyomyces*, and not infrequently a half-eaten specimen may be seen abandoned at the side of the excavation. An extensive study in north American coniferous forests has shown that spores of hypogeous fungi, apparently in a viable condition, are abundant in the faecal pellets of mice, shrews, squirrels and voles.

The fungi developing on the dung of herbivores (see p. 152) are dependent on them for dispersal. Outstanding genera of these coprophilous fungi are *Pilobolus*, *Ascobolus*, *Sordaria* and *Coprinus*. Under natural conditions the sporulation occurs on the dung in a pasture and their spores get on to the grass. A number of these fungi discharge their spores to a distance of 100–1000 mm, sufficient to reach the surrounding herbage without the aid of wind. Further their spore-guns are mostly aligned by positive phototropism. This phototropism is shown by the sporangiophore of *Pilobolus*, by the individual asci of *Ascobolus* and by the perithecial neck of *Sordaria*. The spores, usually associated with mucilage, remain stuck to the grass. In many species during this time of exposure, the spores are protected from the injurious effects of strong light by dark spore-walls (as in *Ascobolus*, *Sordaria* and *Coprinus*) or by the spores being covered by a black sporangial wall (*Pilobolus*). Eventually the grass may be eaten by a herbivore and the spores not only pass uninjured through the alimentary canal, but usually encounter there conditions that stimulate their subsequent germination in the dung.

Man himself has played a part in the dispersal of fungi. Especially significant has been his part in the intercontinental spread of those causing diseases of plants. For example, in the early days of poor communications, the Atlantic Ocean offered an effective barrier to the natural spread of fungi. However, during the past century and a half serious fungal pathogens have crossed the Atlantic Ocean, often as contaminated plant material. Notorious was the introduction of potato blight due to *Phytophthora infestans* to western Europe in the first half of the nineteenth century and the spread of American gooseberry mildew from North America in the early years of the twentieth century. A few decades later the Dutch elm disease, caused by *Ceratocystis ulmi*, spread from Europe in the reverse direction. In the New World a particularly aggressive strain of the parasite developed and this was inadvertently sent back to Britain on infected logs from North America with disastrous results.

So far this account of dispersal has dealt only with terrestrial fungi, but there are aquatic species, although these form a tiny percentage of the whole. Amongst these fungi we may recognize the presumably primitive types with naked zoospores as in *Saprolegnia* and chytrids. Probably the motility of the zoospores is of small account compared with the general movements of the surrounding water. Most likely the biological value of the zoospore relates to the initial stages in dispersal involving escape from the zoosporangium, and to the final stage when the chemotaxis of the motile spore may allow it to 'choose' an appropriate substratum.

As well as primitive, zoosporic fungi in the aquatic habitat, there are numerous species with walled spores both in the sea and in freshwaters. These probably are not primitive aquatics, but are migrants from terrestrial habitats.

Many have spores of an unusual shape which seems to relate to deposition, the final stage in dispersal. In particular it has been argued that the tetraradiate conidia, seen in a number of Hyphomycetes (Fig. 6.5) so common on submerged decaying leaves in streams, represent minute anchors which can make effective contact with the substratum they colonize under conditions of considerable turbulence.

Fungal genetics

8

A study of the biology of fungi would be incomplete without some understanding of the causes of their genetic variability. They are highly variable organisms. Mechanisms to promote outbreeding are important in this respect. The fungi have proved to be ideal organisms for genetical analysis. Some 30 species have been intensely studied and have contributed a great deal to our understanding of genetic systems in general. The reasons for this are manifold.

The vegetative hyphae of the majority of fungi are haploid. For example, those of all Zygomycotina and Ascomycotina and most Deuteromycotina have a prolonged and dominant vegetative haplophase. Mutations arise, are easy to induce and are immediately expressed. There is no system to mask genetic deficiencies, no buffering by heterozygosity or genetic complementation, thus the effects of natural selection are more rigorous and more immediate than in most diploid organisms. Many are usually easy to grow in culture and have a relatively short life-cycle. They often reproduce asexually (by mitosis) and sexually (by nuclear fusion followed by meiosis). By careful manipulation the four separate nuclei produced by meiosis can be separated in the Ascomycotina and Basidiomycotina such that the sequence of gene segregation can be established.

Fungi also have their own unique genetic systems in dikaryosis, heterokaryosis and parasexuality. The existence of hyphal anastomosis or fusion, especially in septate fungi, followed by nuclear migration has already been noted. This ability is of major evolutionary significance in the fungi. It is a key event in the establishment of such systems as it is in many other attributes of the fungi. One property of septate fungi which sets them apart from all other organisms is the ability within a species for cooperation to replace competition. Buller's original example of *Coprinus sterquilinus* illustrates this phenomenon. *C. sterquilinus* is a coprophilous fungus and its spores, after being deposited on grass and subsequently eaten, pass through the gut of herbivores, such as horse, and germinate in the dung. In any one dung ball many may germinate and before long their branched hyphal systems will begin to compete with each other for carbon and nitrogen sources in particular. There may be insufficient of these for each spore to complete its life-cycle and produce a basidiocarp. Hyphal fusions may arise between the hyphae from different spores and the whole becomes one large interconnected mycelium. The whole becomes a single functional unit. Food reserves are pooled and all cooperate to produce one or more basidiocarps. Such an attribute is especially important in the exploitation of substrates containing limited nutrients. This is an important aspect of what Buller termed the 'social organization' of the fungi. It also makes the concept and limits of an individual fungus difficult to define.

Nuclear phases differ amongst fungi. Unlike in other organisms, some possess a dikaryophase. Dikaryosis is where two compatible nuclei form a pair – a dikaryon. They then divide together and the sister nuclei are separated into two daughter cells. Such repeated divisions produce a dikaryophase where each cell contains two haploid nuclei rather than one diploid nucleus. In Ascomycotina, such as *Ascobolus stercorarius* (p. 57), the life-cycle is predominantly haploid, with a vegetative haplophase. The essential part of the sexual process is initiated by the fusion of two cells containing one or more nuclei, a process called plasmogamy. Nuclear fusion or karyogamy does not occur immediately. A dikaryon is established and, by repeated nuclear and cell divisions, a dikaryophase develops. In Ascomycotina this is of limited extent and is purely reproductive in the form of the ascogenous hyphae surrounded by haplophase hyphae of the ascocarp. Nuclear fusion occurs in the ascus initials followed immediately by meiosis (Fig. 4.1). In the Basidiomycotina, it is the haplophase which is restricted. The hyphae produced from two compatible haploid basidiospores soon fuse to establish the dikaryophase (Fig. 5.10). The mycelium has binucleate cells and it is the vegetative phase which is the dikaryophase. Basidiocarps are formed on this and are wholly dikaryotic. Nuclear fusion occurs in the young basidium. Again there is no mitotic divison of the diploid nucleus. Meiosis occurs immediately.

The vegetative dikaryophase is peculiar to the Basidiomycotina. The dikaryophase is functionally diploid, e.g., complementation may occur. The dikaryon may have a critical selective advantage over the diploid in terms of the establishment of dikaryotic mosaics. This involves the direct exchange of one nucleus of the pair between adjacent dikaryons after hyphal fusion to establish a new dikaryon of different genotype. This provides immediate variation in genotype without undergoing the disturbances and delays of meiosis and so permits rapid adaptation to changing environments. This is advantageous in the short term but needs a back-up system, which Basidiomycotina possess, of eventual nuclear fusion and meiosis to provide still further new recombinations of genes producing further variable genotypes.

Many fungi exhibit heterokaryosis. They possess hyphae with at least two but usually more genetically different nuclei in their hyphae. The hyphae are called heterokaryons. Homokaryons possess nuclei of all the same genotype. A dikaryon is a heterokaryon in the sense that its nuclei are of different genotype but there is a major difference, apart from the fact that in the dikaryon the nuclei always divide together such that the ratio in any cell is always 1 : 1. In the dikaryon the nuclei eventually fuse and undergo meiosis, as part of the sexual process. In heterokaryons, this is not usually so; the heterokaryon breaks down.

Heterokaryons arise by hyphal fusion followed by nuclear migration between genetically different mycelia of the same species or by mutations within existing nuclei in a homokaryon. If such events occur in the apical compartment of a hypha, with growth and nuclear division a heterokaryon is established. As most fungi are haploid in the vegetative state heterokaryosis confers the genetic advantage normally associated with diploidy and heterozygosity. A dominant wild type allele for a particular character would mask a recessive mutant allele at the same gene locus in another nucleus. For example, biochemical mutants of *Neurospora crassa*, one lacking the ability to synthesize *p*-aminobenzoic acid and the other nicotinic acid, fail to grow on minimal media lacking these ingredients. In *N. crassa*, the genes controlling the synthesis of these two substances are non-allelic. The establishment of a heterokaryon between the two mutant types would mean that both nuclei would exist in the same cytoplasm and each mutant would complement the deficiency of the other. The observation that heterokaryons make better growth than their component homokaryons has often been noted. In addition the relative proportions of the various genetically different nuclei can vary within the hyphae and can be altered by environmental forces. The rates of multiplication of the different kinds of nuclei may be differentially altered such that the composition of the heterokaryon is changed. This confers the advantage of rapidly responding to selection pressures such as changing nutrient or other environmental regimes.

Any heterokaryotic mycelium may give rise to homokaryotic tips. This often leads to sectoring of colonies in culture and can be seen in *Botrytis cinerea*. Each conidium of *B. cinerea* contains several nuclei and, if these are not genetically alike, a heterokaryotic mycelium results from its germination. By chance as the colony develops, a leading hypha may contain only one of the nuclear types present, and so give rise to a homokaryotic sector of the colony with its nature determined by the genotype of that nucleus. The sector may be distinguished, for example, by producing more or fewer conidiophores (Fig. 8.1). The chance production of homokaryotic hyphal tips is a common way in which heterokaryons break down. Another way is in the formation of uninucleate spores.

Although in *B. cinerea* conidia are multinucleate, in many fungi they are uninucleate. This is so in *Aspergillus nidulans* which has been used extensively in the study of heterokaryosis and related phenomena. The fungus is really an Ascomycotina and its ascocarpic state, a cleistothecium, is readily produced in culture. The anamorph is of the characteristic *Aspergillus* type. The conidia, formed in chains from a single head, are uninucleate. A number of strains of *A. nidulans* are known. In one, for example, the colony is yellow and another green, the colour being in the conidia. By growing these strains together a heterokaryon can be produced. In this some conidiophores produce at their apex separate chains of both yellow and green coloured conidia. The swellings from which the phialides, the spore-producing cells, arise may contain both types of nuclei involved in the heterokaryon. Each phialide however is

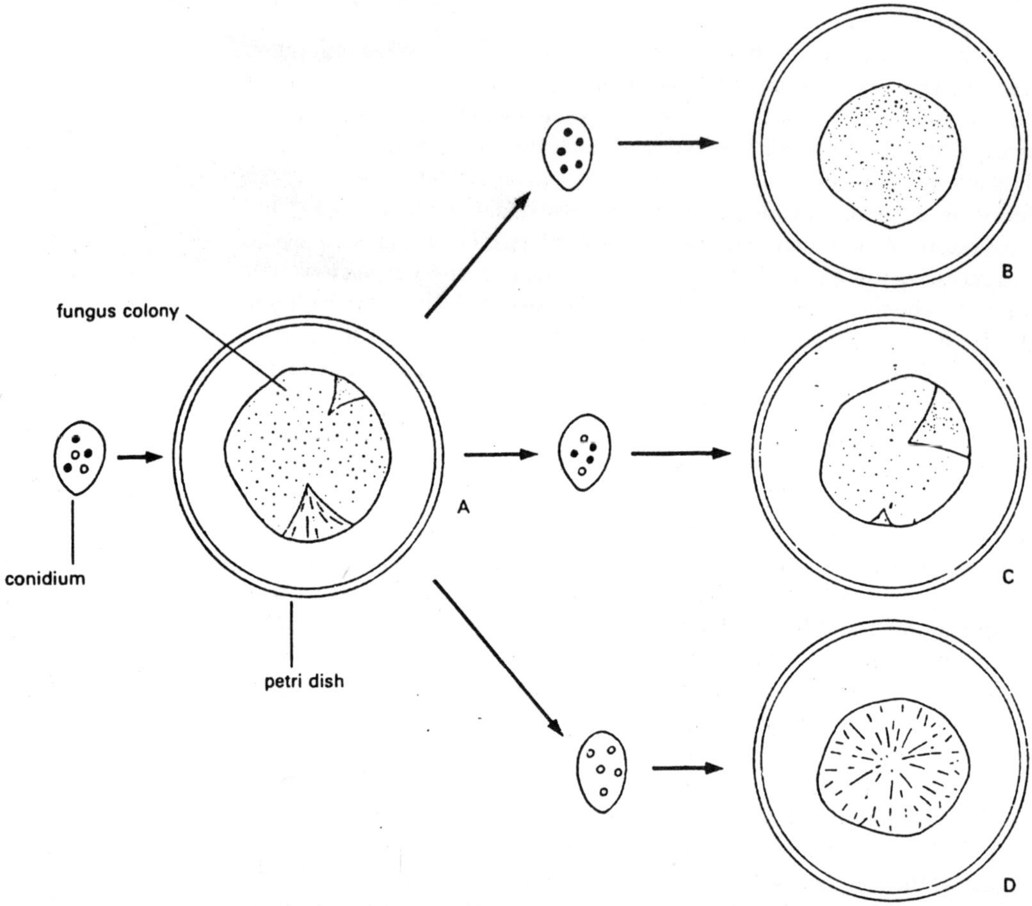

Fig. 8.1. Diagram of heterokaryosis in *Botrytis cinerea*. A, colony derived from a single conidium producing a poorly-sporing sector (above) and a richly-sporing one (below). A conidium from each sector gives non-sectoring colonies (B and D) but a conidium from the main part of the colony gives a colony sectoring like the parent one (C). The supposedly different types of nuclei in the conidia are represented by black and white dots.

uninucleate and so produces either a chain of yellow conidia or a chain of green ones (Fig. 8.2).

Heterokaryosis confers considerable genetic plasticity on a mycelium and provides a mechanism for the recombination and segregation of its component whole nuclei. How important it is in nature is still a matter for questioning. They are rarely isolated from nature. In synthesizing heterokaryons they are only readily formed between strains with closely similar genetic backgrounds. Vegetative incompatibility systems may restrict or even prevent heterokaryon formation between genetically very dissimilar strains. This might restrict its potential in providing variation in nature. It fails in the sense that it breaks

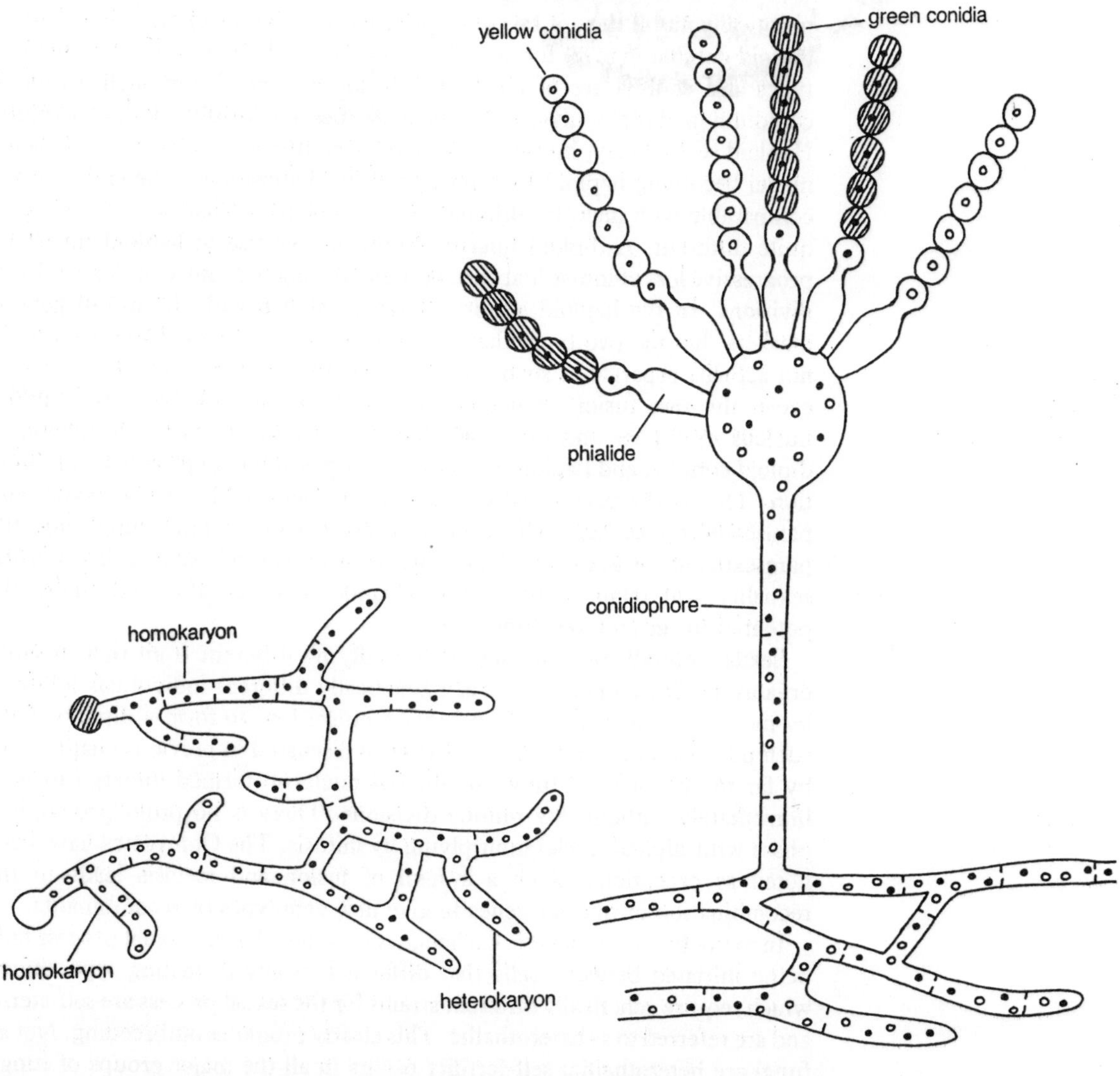

Fig. 8.2. *Aspergillus*. Diagram of a heterokaryotic mycelium (heterokaryon) with two genetically different types of nuclei. The conidia produced are uninucleate and therefore homokaryotic. On germination they produce homokaryons which may unite to give a heterokaryon.

down at asexual spore formation in many fungi and in that nuclear fusion followed by meiosis does not occur to exploit genetic recombination. Each nucleus usually retains a genetic entity however much they are mixed.

Heterokaryosis is very common in Deuteromycotina and many of these are extremely variable. One explanation is that there has been a form of recombination between nuclei in these via a phenomenon known as parasexuality. In *Aspergillus nidulans*, very occasionally two nuclei of different genotype may fuse in the heterokaryon to form a diploid one. This divides mitotically and if this, or one of its progeny, is found in a phialide, a chain of diploid conidia may be formed. This differs in colour from the two haploid types and conidia are of about double the volume. From such a diploid conidium a diploid mycelium can grow bearing mainly diploid conidia. Haploid sectors may, however, develop apparently as a consequence of some nuclei becoming haploid by a process of 'haploidization'. The end result is comparable with meiosis, although the cytological mechanism appears to be quite different. A diploid nucleus seems to give rise to haploid nuclei by progressive loss of individual chromosomes during the course of many mitotic divisions. In the haploid sectors, it can be shown with the use of genetic markers that the two types that contributed to the original heterokaryon do not actually separate as such, for recombination occurs. Thus there exists a cycle: the rare fusion of haploid nuclei in the heterokaryon; the diploid nucleus multiplies and may lead, but not necessarily, to the formation of diploid conidia; and haploidization with the possibility of genetic recombination. This is the parasexual cycle. In *A. nidulans* both normal sexual and parasexual cycles exist. However, in a great many conidial fungi only the parasexual one is available. Thus fungi which seem to have lost their normal sexuality still retain some of the advantages of sexuality, especially the potential for genetic recombination.

Sexual reproduction in fungi is basically no different from that in other organisms. Sexual fusion between genetically different individuals followed by meiosis is widespread. Two haploid nuclei fuse to form a diploid. How such nuclei are brought together has been discussed in previous chapters. In by far the majority of fungi the diploid nuclei so formed undergo meiosis immediately without any mitotic divisions. There is no prolonged diploid phase with diploid nuclei multiplying by mitosis. The Oomycetes have been noted as exceptions. Such a system of fusion and meiosis leads to the recombination of nuclear genes to give new genotypes or recombinants.

In many fungi this genetic exchange is promoted by the sexual process only being initiated between cells that differ genetically in mating type. Fungi which require genetically different strains for the sexual process are self-sterile and are referred to as heterothallic. This clearly promotes outbreeding. Not all fungi are heterothallic; self-fertility occurs in all the major groups of fungi. This homothallism does not preclude outbreeding but reduces the likelihood of it occurring. In many Mucorales, zygospores are formed in culture from single, uninucleate sporangiospores, but in most species of *Mucor* cultures derived from a single sporangiospore fail to form zygospores. This problem was investigated by Blakeslee in the early years of this century. His researches form a landmark in the study of fungi, opening up the immense field of fungal genetics. Blakeslee found that if, in a particular species of *Mucor*, a large number of single spore-spore cultures were obtained from different sources, they could be sorted into two types or strains, which he called 'plus' and 'minus'. If in a culture dish a 'plus' colony met another 'plus' one, or if a 'minus' encountered another 'minus', there was no sexual response. However, if a 'plus' and 'minus' met, zygospores were formed in the zone of contact. The 'plus' and 'minus' strains were morphologically alike. It was not possible to designate one strain male and the other female. The two strains are said to differ in mating type. A species with 'plus' and 'minus' strains is said to be heterothallic, and the phenomenon known as heterothallism. The genetical basis seems to be that mating is determined by a single gene locus with two alleles, 'plus' and 'minus'. These mating type loci may be variously designated '+' and '−', 'A' and 'a'. They determine the ability of a strain to cross sexually with another strain.

It is now clear that in many fungi the processes leading up to plasmogamy and karyogamy in sexual reproduction are highly complex and regulated by diffusible hormones or pheromones. The two best known systems are those in *Mucor* and *Achyla*. Interactions between 'plus' and 'minus' strains have been the subject of much research. In *Mucor*, the first visible effect seen when 'plus' and 'minus' strains are grown opposite to each other in a petri dish is the production of aerial hyphae, zygophores. These are distinguished from ordinary hyphae in being wider, smoother and with their tips containing more yellow carotenoids. Their induction is triggered by the build-up of active hormones, trisporic acids. Neither strain alone can produce these but both produce different hormone precursors or inducers, Prohormone P^+ and Prohormone P^-. These diffuse through the agar and each is converted by specific enzymes from the other strain to trisporic acids. These prohormones are structurally very similar to trisporic acids and can be readily converted into these. Subsequently the zygophores show directional growth or zygotropism towards each other. It is thought that another hormone, zygotropin, produced by the zygophores is responsible for this, but its identity has not been determined. Finally gametangia are delimited, probably stimulated by still further hormones. In *Phycomyces blakesleeanus*, as in *Mucor*, hormones are involved in the initiation of the sexual process but aerial zygophores are not formed. Instead, as colonies of opposite strain approach, the hormone from one strain diffusing through the agar causes certain submerged hyphae of the opposite strain to swell. Such swollen hyphae of opposite strains grasp each other to form a knot-like structure within the agar. Still adhering they grow upwards together, but on emerging from the agar surface their tips arch apart as the progametangia (Fig. 3.11).

It is to be noticed that similar systems involving chemical coordination of the sexual process are found elsewhere in the fungi. In some Saprolegniales there is a sequence of processes, involving the formation of gametangial branches, chemotropism and delimitation of gametes, that requires two and perhaps more hormones. For example, *Achyla* is heterothallic and strains involved in mating respond by producing only one kind of gametangia, antheridia or oogonia. When grown by themselves gametangia are not differentiated. When two compatible strains are juxtaposed, the leading hyphae of the antheridial strain shows inhibition of growth and branches profusely near the tip. These branches are much narrower, more irregular and much more branched than normal branches. These are the antheridial hyphae and their formation is induced by the hormone antheridiol, produced at all times by female strains. Antheridiol also stimulates the antheridial hyphae to produce the hormone oogoniol, which stimulates the female strain to produce oogonial initials. Antheridiol may also cause chemotropic growth of the antheridial branches. It appears that oogoniol, or another hormone, produced by the antheridia may cause differentiation of eggs in the oogonia.

In Ascomycotina, such as *Ascobolus stercorarius*, cultures derived from single ascospores fail to form apothecia. If a large number of such cultures are produced, these can, as in *Mucor*, be sorted into 'plus' and 'minus' strains alike in appearance. In *Ascobolus*, however, both strains are hermaphrodite. When 'plus' and 'minus' strains are grown together apothecia are produced. This is the general situation in heterothallic Ascomycotina and is identical with that found in the Zygomycotina and Oomycetes. The two strains are usually designated 'A' and 'a' which represent two alleles at a single gene locus determining the ability to mate. The alleles 'A' and 'a' segregate during meiosis in the ascus. Segregation of alleles of other genes also occurs and their pattern of segregation in an ascus is more easily detectable. There is a mutant strain of *Ascobolus* producing colourless spores. In each ascus of hybrid apothecia, in a cross between this and a normal strain, there are four purple and four colourless spores. There is now a great body of genetical work on Ascomycotina mostly based on the perithecial fungus *Neurospora* (similar to *Sordaria*, p. 61) in which the ascospores form a single row in the ascus. The geneticist thus has an 'ordered tetrad'. Each spore of an ascus can have its position noted, be removed from the ascus by micromanipulation and separately cultured. From such studies it can be determined whether segregation of the alleles of a gene has occurred at the first or at the second meiotic division. Because the final division is mitotic, contiguous sister ascospores have the same genetical constitution. There are six possible patterns (Fig. 8.3), two involving separation of the alleles at the first division of meiosis and the other four at the second.

In the Zygomycotina the situation, as we have seen, is not always so clear cut (Chapter 3). In *Mucor*, for example, the haploid 'plus' and 'minus' nuclei are brought together at zygospore formation. They probably fuse pairwise giving a number of diploid nuclei, although some nuclei may remain unpaired. It is generally agreed that meiosis occurs at some stage during zygospore germination. We have noted that on germination of the zygospore a normal sporangium, known as the germ-sporangium, is formed from this (Fig. 3.14). Analysis of the spores from this sporangium shows that they are all 'plus' or all 'minus'. In an attempt to explain this, it has been suggested that in *Mucor* meiosis is confined to a single surviving nucleus, all the others degenerating. It is further proposed that following meiosis, when the two 'plus' and 'minus' alleles segregate, only one of the four haploid nuclei survives (Fig. 3.15).

More complex compatibility systems occur in the Basidiomycotina. A variation on the above system is to have multiple alleles at the locus determining the ability to mate. In addition there may be two gene loci involved, A and B, with either or both having multiple alleles. Heterothallism where two genes are involved is known as tetrapolar in contrast to bipolar type involving only one gene determining mating type, as found in Ascomycotina and Zygomycotina. Tetrapolar heterothallism is the commonest type in the

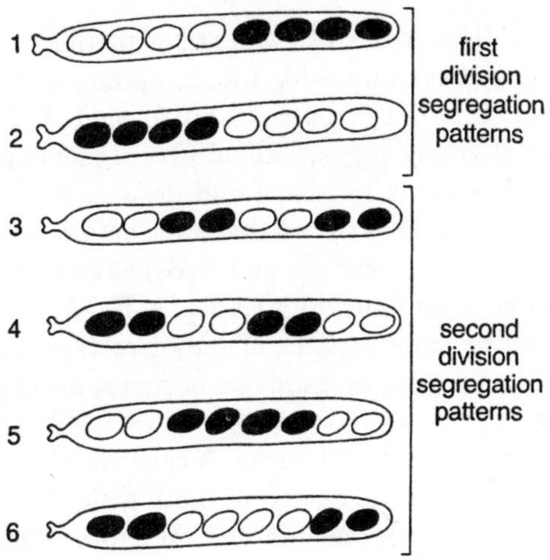

Fig. 8.3. *Neurospora*. The six segregation patterns amongst asci in a perithecium. These would be directly visible if the perithecium were a hybrid one produced by mating a black- with a white-spored strain. The same patterns result when mating-type alleles (A and a) are concerned. Types 1 and 2 are formed in equal numbers. Types 3, 4, 5 and 6 are also formed in equal numbers. The ratio of types 1 and 2 to the other types depends on how near is the gene on the chromosome to the centrometre.

larger Basidiomycotina, although bipolar species are not infrequent. A few, some 10%, are homothallic.

In agarics, such as *Coprinus comatus*, the shaggy ink-cap, and *Piptoporus betulinus*, the birch polypore, when mycelia grown from single basidiospores from a single basidiocarp are paired, fertile dikaryons are formed in only half of the crosses. This is identical with the situation in Ascomycotina. But when mycelia grown from single basidiospores from basidiocarps collected over a wide area are paired, fertile dikaryons are formed from virtually all pairings. This can be explained by the fact that there are multiple alleles at the mating type locus. Thus the A gene locus has alleles A_1, A_2, A_3, A_4 . . ., etc. In *Coprinus cinereus*, and in most other heterothallic agarics, when mycelia grown from single basidiospores from a single basidiocarp are paired, fertile dikaryons result in only one-quarter of the pairings. The mating type is determined by two genes A and B which occur on different chromosomes as indicated by their independent assortment at meiosis. In any basidiocarp, each gene is represented by two alleles, for example A_1 and A_2, and B_1 and B_2. The four mating types, with the resulting dikaryons would be as follows:

monokaryon A_1B_1 — $(A_1B_1 + A_2B_2)$ dikaryon
monokaryon A_2B_2 —

monokaryon A_1B_2 — $(A_1B_2 + A_2B_1)$ dikaryon
monokaryon A_2B_1 —

A dikaryon is only formed between two monokaryons if the two alleles of both genes are different. Both pairs of compatible matings produce a diploid nucleus in the young basidium with the composition $A_1A_2B_1B_2$. Meiosis determines that four kinds of basidiospores are formed; A_1B_1, A_2B_2, A_2B_1 and A_1B_2, each capable of giving a genetically different monokaryotic mycelium.

It has been found possible to isolate and grow each spore from a single basidium. Work of this kind has shown that some basidia produce two types of spore (either A_1B_1 and A_2B_2, or A_2B_1 and A_1B_2) and two of each type; while others produce spores of all four kinds (A_1B_1, A_2B_1, A_1B_2, A_2B_2) (Fig. 8.4). Whether a basidium produces spores of only two types, or bears all four, depends on whether separation of the alleles occurs at the first or at the second division of meiosis.

Multiple alleles may occur at each locus. A particular basidiocarp may have as alleles of the A gene, A_1 and A_2; and of the B gene, B_1 and B_2. However, another basidiocarp from some way off, and therefore clearly derived from a different mycelium, may carry A_3 and A_4, together with B_3 and B_4. Any monokaryon from the first basidiocarp (A_1B_1, A_1B_2, A_2B_1 or A_2B_2) can mate with any from the second (A_3B_3, A_3B_4, A_4B_3 or A_4B_4). This is in agreement with the rule already stated that for mating to occur the alleles of both genes must be different in the two monokaryons.

Both mating-type genes have multiple alleles. Thus the A gene has alleles A_1, A_2, A_3, A_4 . . ., etc., and similarly for the B gene. There appear to be about 160 alleles of A, and about half that number of B. However, in any one dikaryotic mycelium, and in the basidiocarps derived from it, only two A alleles and two B alleles can coexist.

In connection with the change from the monokaryotic to the dikaryotic state, a remarkable phenomenon should be mentioned. It is known as the Buller phenomenon, after the late A.H.R. Buller whose contributions to an understanding of the biology of fungi were outstanding. If a petri dish with a largish monokaryotic colony (say A_1B_1) is inoculated with a compatible type (say A_2B_2) so that a tiny colony of the latter forms close to the larger one, they soon meet and vegetative anastomoses occur. Then, in the course of a day or two, the developing edge of the large colony becomes dikaryotized producing mycelium with clamp connections. What appears to happen is that A_2B_2 nuclei enter the A_1B_1 mycelium and migrate through it, probably dividing in transit. Thus the large colony is dikaryotized, and so, incidentally, is the small one. The radial growth of the mycelium in *Coprinus cinereus* is 2–3 mm a day, but the speed of migration of the nuclei is around 20 mm a day.

Currently much attention is being paid to the molecular structure and function of the mating type genes. In a number of Basidiomycotina the situation is far more complex than first thought and it appears that each allele at the A and B loci are really pairs of closely linked functional genes. The term 'allele' might not be the most appropriate one to use and 'idiomorphs' has been suggested as a substitute. As clusters of functionally related genes they fit the concept of 'supergenes'. In Basidiomycotina the function of cell fusion leading to dikaryon formation is ascribed to the A locus and nuclear division and migration and sexual development to the B locus, but again this may be oversimplistic. In *Ustilago maydis*, which causes maize smut, the A locus has

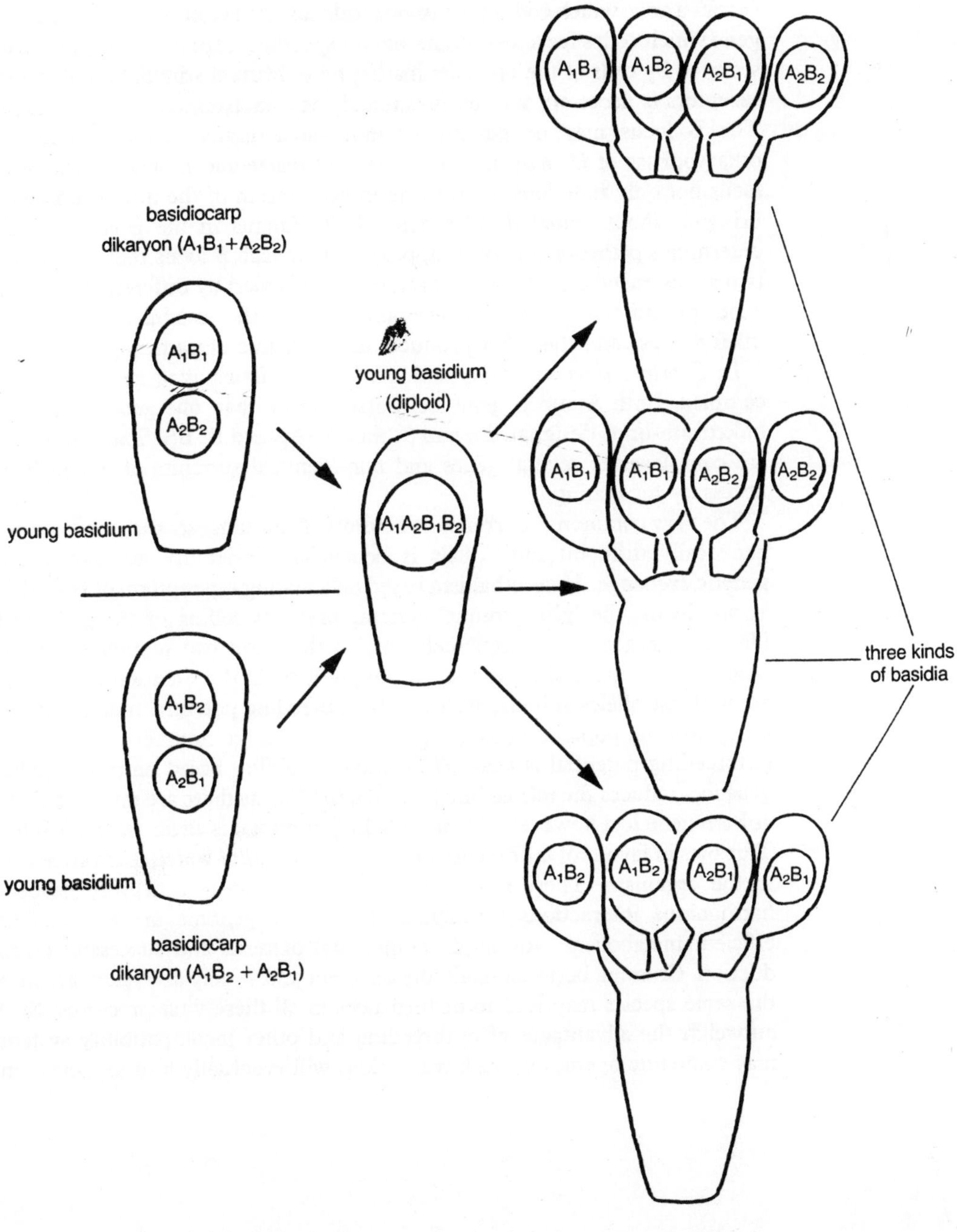

Fig. 8.4. Diagram of genetical behaviour in the basidium. All young basidia in a basidiocarp have the same type of diploid nucleus ($A_1A_2B_1B_2$) but there are three types of basidia.

two alleles and the B locus 25 alleles, but both of the loci appear again to consist of at least two functional linked genes. Each of the so-called A alleles carries a gene which codes for a lipopeptide mating factor or pheromone and a gene which codes for a membrane-bound specific receptor for the pheromone secreted by cells of the opposite mating type. Mutual stimulation of the two receptors is necessary for cell fusion and the establishment of the dikaryon. The B locus may be equally or more structurally complex. Only the dikaryophase of *U. maydis* is able to infect maize and it appears that the B locus not only functions in ensuring multiplication of the dikaryophase and bringing about sexual development of the fungus in the host but it also determines pathenogenicity. It appears that at each B locus there are at least two genes encoding regulatory proteins. If specified by different alleles, one gene programmes sexual development after the establishment of the dikaryophase and the other produces determinants of pathenogenicity.

In *Coprinus cinereus,* the situation with the mating type genes is equally complex. Both A and B gene loci possess more than one gene. They have linked sub-units designated α and β, thus AαAβ and BαBβ. The α and β loci are complexes of several genes and non-identical subunits of A and B are necessary for mating.

The very common occurrence of heterothallism suggests that the mating of genetically different individuals is beneficial. Variability is promoted by genetic exchange. Heterothallism based on a mating type system of two alleles at one locus, the 'plus'/'minus' system, prevents selfing of the genetically identical progeny of a single cell but, as there are two mating types, the potential for outbreeding and inbreeding is 50%. In the situation where there are multiple alleles at one gene locus the inbreeding potential remains at 50% but, since the population as a whole contains a large number of alleles, the outbreeding potential is near 100%. Incompatibility based on two unlinked gene loci reduces the inbreeding potential to 25%; as there are multiple alleles at both gene loci however, the outbreeding potential is again virtually 100%. This bias in favour of outbreeding promotes variability which is advantageous to the organism. There is no doubt that this is so to a degree, although harmonious interactions throughout the whole genome are essential for efficient metabolism, normal developmental patterns and successful reproduction. Crossing between markedly different and widely divergent strains of the same species may lead to disturbances in all these vital processes. Such outweigh the advantages of outbreeding and other incompatibility systems may come into operation. Such restrictions will eventually lead to speciation.

Ecology of saprotrophic fungi 9

In the preceding chapters fungi have been introduced in a taxonomic framework with a few digressions on matters of biological interest. Many fungi have adopted a saprotrophic mode of nutrition and in this mode they excel as decomposers. Here it is proposed to look at some of the more important aspects of the ecology of saprotrophic fungi.

The most significant activity of such fungi is their special ability to break down cellulose. About one-third of all the organic matter produced by green plants is in the form of cellulose. Cereal straw, for example, consists of about 40% cellulose. Much of this cellulose is locked up in woody tissues, especially in trees. The bulk of woody tissues consist of dead and empty xylem vessels, tracheids or fibres. Such wood is composed of three major components – 40–60% cellulose, 10–30% hemicelluloses and 15–30% lignin. Whereas many saprotrophic fungi can utilize cellulose and hemicelluloses, there are far fewer, mainly Basidiomycotina and a small number of Ascomycotina, that can attack lignin and so decompose wood. Such fungi perform a vital and useful function in both land and aquatic habitats in recycling carbon, and minerals, temporarily out of circulation in these complex organic compounds.

It should be stressed that the fungal mycelium, with its battery of enzymes secreted by the growing hyphal tips and aided by the hydrostatic mechanical thrust of the turgid extending tips, is uniquely fitted to attack a bulky fibrous structure such as the trunk of a tree.

To what extent wood resists fungal attack depends largely upon its moisture content. Dry wood is not liable to invasion. It cannot be colonized unless the moisture content is above 26–32% on a dry mass basis depending upon the wood. Air-seasoned worked wood contains 15–18% moisture which is far too low to support fungal growth. Completely waterlogged wood is immune to fungal attack mainly because the oxygen tension is too low to support hyphal growth. Of all the wood decay fungi *Serpula lacrymans*, the dry rot fungus, is the exception (Chapter 12). Once established, on a small pocket of damp wood, it produces metabolic water from the degradation of cellulose which raises the moisture content of the wood on which it is growing and is thus able to colonize even dry timbers in this way. Two other features contributing to the resistance of wood to decay are its very low nitrogen content – 0.03–1.0%,

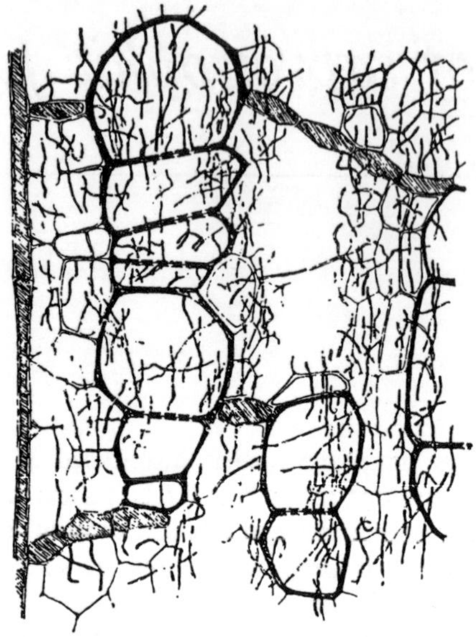

Fig. 9.1. Wood rot. Small part of a transverse section of beech (*Fagus*) wood in late stage of rotting by *Ustulina deusta*. Many of the wood fibres have been completely destroyed and the walls of others rendered very thin. The wood vessels have retained their form. Fungal hyphae pass from element to element through the pits. (After Wilkins)

compared with 1.0–5.0% in herbaceous tissues, and the deposition of toxic substances, mainly phenolics, during the formation of the heartwood. The carbon : nitrogen ratio in most woody tissues is in the order of 350–500 : 1 or even higher. For the majority of fungi a substrate with such a high carbon : nitrogen ratio would be nitrogen deficient and growth limiting. Toxic substances are synthesized in the senescing wood parenchyma cells and diffuse out into the walls of the adjacent xylem vessels and tracheids as the heartwood is formed. They fall into a number of chemical groups but of these tropolones, such as thujaplicin, are the most inhibitory. They all provide protection from decay for a number of years but with time they may be leached out or inactivated. In spite of their toxicity several fungi are able to destroy the heartwood of even living trees, and even timber impregnated with chemicals, such as pentachlorophenol, which are widely used to protect less durable timbers. A durable heartwood may be of survival value to the tree itself. Some cedars have very resistant heartwood and may live 2000 years or more. Nevertheless, normally dead trees and stumps in forests are quickly colonized by fungi.

Two principal types of rot can occur. First there are the white rots, such as that caused by *Coriolus vesicolor*, one of the commonest bracket fungi. *Ustulina deusta*, an abundant pyrenomycete, is responsible for a white rot of beech (Fig. 9.1). In such rots, the wall polysaccharides, such as cellulose and hemicelluloses, are attacked more or less simultaneously as the lignin and the wood becomes markedly paler, and more fibrous as the pigmented amorphous lignin is removed. Secondly there are brown rots, such as that caused by *Piptoporus betulinus* in birch (*Betula*). Here cellulose and hemicellulose are used leaving the lignin virtually unaffected with the result that the wood becomes a darker brown colour. Whereas in white rots there is a general thinning of the cell walls of the xylem as all the wall components are used, in brown rots there is no thinning of the walls. The structural polymers are removed leaving a framework of lignin to maintain the general cell shape. Decomposition often also occurs in irregular patches in the attacked wood. This leads to a cubically cracked appearance in the rotted wood. It is also no longer fibrous and readily crumbles into a powder when rubbed between the fingers.

Several hundred species of agarics and polypores, especially the latter which are almost entirely confined to growing on wood, cause white rots. *C. versicolor* is one of the most efficient. It can degrade over 90% of the lignin in most woods. Apart from these only a few Ascomycotina, such as *U. deusta* and *Xylaria polymorpha* can cause a white rot. Somewhat fewer Basidiomycotina cause brown rots. Thus in the fungi as a whole, the ability to degrade lignin completely is limited to the relative few.

Dead trunks, stumps and fallen branches of beech (*Fagus*) support a particularly large variety of saprotrophic fungi. Of the numerous agarics which may be present some, such as *Oudemansiella mucida* and *Pholiota adiposa*, have a very strong preference for beech, and the same is true of the polypores *Daedaleopsis confragosa* and *Bjerkandera adusta*. Others, such as the agaric *Hypholoma fasciculare*, the sulphur tuft fungus, and the bracket fungi *C. versicolor* and *Stereum hirsutum*, although abundant on beech are regularly to be found on most kinds of wood. The agaric *Oudemansiella radicata* is abundant in beech woods. It appears to be growing in the leaf-litter, however, if its base is carefully excavated, it can be seen that the stalk is prolonged downwards some 100–200 mm and is then found to have arisen from a dead root or buried branch of beech. Many Ascomycotina also occur on beech. Mention may be made of *Hypoxylon fragiforme*, a stromatal type like a *Daldinia* (p. 64), but brick red. It occurs in regiments on dead branches. *Xylaria hypoxylon*, the candle snuff fungus, is also common on cut stumps. Some moulds are also common. *Bispora monilioides* is conspicuous on the surfaces of cut stumps, forming black, tar-like splashes. A beech stump takes several years to rot away, mainly through the activity of many different fungi. In the last stages, when the wood is becoming crumbly, *Lycoperdon pyriforme* is often to be seen. It is unusual amongst puff-balls in growing on wood. It tends to occur in troops arising from white mycelial strands which ramify through the soft substratum.

Turning to oak (*Quercus*) many fewer species are encountered. However, *Daedalea quercina* is a bracket fungus almost limited to oak and so too is the little agaric *Mycena inclinata* which occurs in tufts on the stumps. *Fistulina hepatica*, the beef-steak fungus, a large, soft-textured polypore which when cut looks like lean meat, is to be seen, often high up, on living oak trees. Its mycelium is in the heartwood, causing, however, no loss of mechanical properties. Indeed heartwood of oak invaded by *Fistulina* is much prized by cabinet makers for its rich brown colour. It is one of the few fungi which can tolerate the toxic tannins laid down in the dead heartwood. This undoubtedly accounts for its marked specificity. Another common species on fallen trunks and branches is the discomycete *Neobulgaria inquinans* the black apothecia of which have the consistency of india-rubber.

Again coniferous wood has its own distinctive saprotrophic fungi. On pine stumps *Tricholomopsis rutilans* and *Paxillus atrotomentosus* are characteristic agarics. At the base, often arising from the roots, a large, brown polypore, *Phaeolus schweinitzii*, is frequently seen and, less commonly, the big cauliflower-like basidiocarps of *Sparassis crispa*. Another species abundant on pine stumps is the orange gelatinous fungus *Calocera viscosa*. This is branched and superficially like a *Clavaria*, but belongs to quite a different order of Basidiomycotina.

Elms (*Ulmus*) killed by Dutch elm disease call for special mention. The fungal pathogen (*Ceratocystis ulmi*) is not a wood-rot fungus; it simply poisons the tree leaving the wood intact. However, once dead, the tree is invaded by saprotrophic fungi which attack the substance of the wood. There is a fairly regular succession. Within a year or two of death, the tree is usually colonized by *Flammulina velutipes* (p. 89), but after a few seasons this tends to be displaced by *Pleurotus cornucopiae*. Five or six years later, when usually the dead trees have been felled, the jelly fungus *Auricularia mesenterica* takes over the stumps. Another fungus frequently to be seen on elm is *Polyporus squamosus*, dryad's saddle. Before Dutch elm disease became epidemic, *P. squamosus* was the most serious threat to the tree. It is a wound parasite which attacks the heartwood, weakening it greatly with the result that major limbs fall off and even the whole tree may topple. The large, saddle-like basidiocarps are soft for a polypore and the tubes on the underside are quite short and rather wide.

The larger fungi on tree stumps are not all harmless. This is particularly true of *Armillaria mellea*, the honey fungus, and *Heterobasidion annosum*, which causes a heart rot of pines (Chapter 10). They are both devastating parasites. Honey fungus is often responsible for the death of mature trees, such as those of beech, and continues to grow on the dead tree many years after its death.

Each tree species thus may have its own particular wood decay fungi, although some of these are less fastidious than others. Specificity may be determined by the nature of the toxic substances laid down in the heartwood. Only fungi which can degrade, tolerate or inactivate these are able to become established. A number of these fungi may gain access as necrotrophic parasites at wounds above ground or along roots and persist as saprotrophs after death of their host. Being established first, they would have a competitive advantage over purely saprotrophic species.

We may now turn to the fungi of the forest floor. Leaf-litter forms an ideal medium for fungal growth, being rich in organic matter and well-aerated. It attracts a wealth of both microscopic and macroscopic species. Even before they fall, leaves on the tree support a variety of 'phylloplane inhabitants'

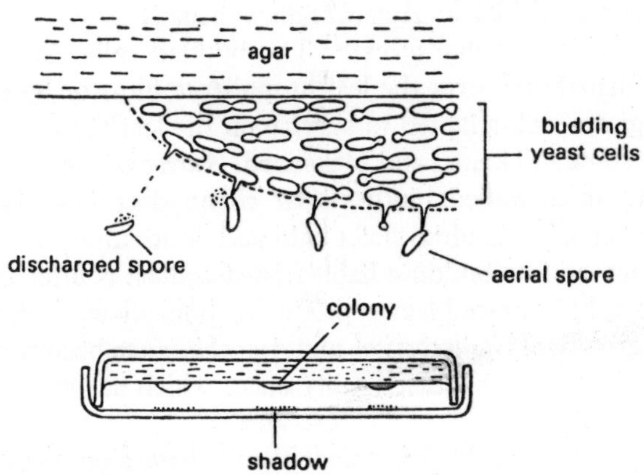

Fig. 9.2. *Sporobolomyces.* Below, vertical section of an inverted petri-dish culture with three yeast colonies and a 'mirror picture' or 'shadow' of each picked out by discharged spores. Above, portion of a colony highly magnified showing yeast cells budding in a slimy colony and others, near the surface, producing sterigmata bearing aerial spores.

existing on organic substances, such as simple sugars and organic amino-acids, which exude from the leaf cells, especially as they age. These include normal yeasts and yeast-like fungi, in particular the mirror-image or shadow yeasts belonging to the family Sporobolomycetaceae. If, for example, a senescent leaflet of bramble (*Rubus fruticosus*) is stuck with vaseline to the inside of a petri dish lid above a surface of sterile agar, tiny yeast colonies develop on the agar below in a few days. They may be so numerous as to form a rough reproduction of the shape of the leaflet above. Most of the colonies are pink. If one is examined directly under the microscope, it is seen to consist mostly of budding cells, as in *Saccharomyces*, but a few of the cells behave differently. From each of these an aerial sterigma is produced bearing a single, aerial spore asymmetrically. This is violently discharged in just the same way as is the basidiospore of a toadstool. Because such behaviour is considered unique to Basidiomycotina, it is generally believed that these yeasts belong to that group. These 'mirror-picture' or 'shadow' yeasts are amongst the most abundant fungi in country districts and their spores contribute enormously to the air-spora (p. 125), particularly in the early hours of the morning. The commonest genus is *Sporobolomyces*. When colonies of this are grown on nutrient agar in an inverted petri dish, discharged spores form an exact mirror-picture of these colonies on the inside of the lid of the dish (Fig. 9.2).

Sporobolomyces roseus is the most common species and is virtually omnipresent on leaves of most plants. It completes its life-cycle on the leaf surface. Other yeasts, such as those in the genus *Cryptoccocus*, exist in the budding phase. Two conidial Ascomycotina *Aureobasidium pullulans* and several species of *Cladosporium* are also phylloplane inhabitants. After impaction on the leaf surface their conidia develop into hyphae forming quite extensive, if sparse, colonies. *Aureobasidium* usually grows by yeast-like budding with minimal hyphal growth. The budded cells, as in the other yeasts, may be distributed over the leaf in moisture films or by rain splash. Both these fungi are well-adapted to survive on the leaf surface. Their cell-walls rapidly become thickened and melanized. *Aureobasidium* also produces clumps of dark, thick-walled multicellular chlamydospores. This enables them to survive exposure to ultra-violet light and desiccation. Because of this dark pigmentation and its budding habit *Aureobasidium* is often classed as a black yeast. On aphid-infested leaves producing honeydew, such as those of lime (*Tilia*) the growth of *Cladosporium* and *Aureobasidium* becomes so intense that they cover the leaves as a black sooty mass, which has earned them the name of 'sooty moulds'.

Thus the leaves falling to the forest floor already have a population of fungi and these may initiate decay. However, soon others replace them. In recent years there has been intensive study of decay of some of the major types contributing to the litter, especially pine needles and oak leaves. However, the process is complex since other organisms, especially small invertebrates and bacteria, are also involved in the decomposition. Decay eventually leads to the production, in the soil below the litter, of amorphous humus derived from complex organic substances relatively resistant to decay such as lignin, chitin and keratin. Although the earlier colonizers of fallen leaves are microfungi, after a time macrofungi, especially agarics, begin to play a part. Common toadstools existing on the leaf-litter are *Laccaria laccata*, *Collybia peronata* (woolly-foot) and, in beech woods, *Mycena pura*.

Beech, oak, pine and other types of woodland tend to have their own characteristic agarics producing their basidiocarps in the litter. In part this is due to the preference of particular species for a particular kind of leaf-litter but in the main these are not litter decomposers but ectomycorrhizal fungi (Chapter 11). The major distinction between these two groups of fungi is in their mode of nutrition. The litter decomposers are good competitive saprotrophs with marked cellulolytic ability and some can degrade lignin. The majority of mycorrhizal agarics have lost the ability to utilize these polymers and rely upon their tree host for organic carbon supplies and exist as obligate biotrophs.

A special habitat encountered in a woodland is the site of a brushwood fire. A small and rather diverse, but distinctive, group of fungi are known to be associated with such fire sites. There are a number of agarics which in nature are found on burnt ground and only rarely elsewhere, especially *Pholiota carbonaria* and *Myxomphalia maura*. Further, many Discomycetes are found on wood-fire sites, especially *Pyronema confluens* forming tiny, pink apothecia in dense crowds and *Geopyxis carbonaria* forming cup-shaped, orange apothecia in discrete clumps. The large apothecia of *Rhizina undulata* also occur in this habitat. It is a parasite of pine, the roots being infected in the neighbourhood of brush-fire sites. In this fungus a temperature of 35–45°C triggers the germination of the ascospores, so that the new mycelium tends to develop and bring about infection of pine roots close to the area where there has been a wood fire. It is not known why fungi are restricted to such sites.

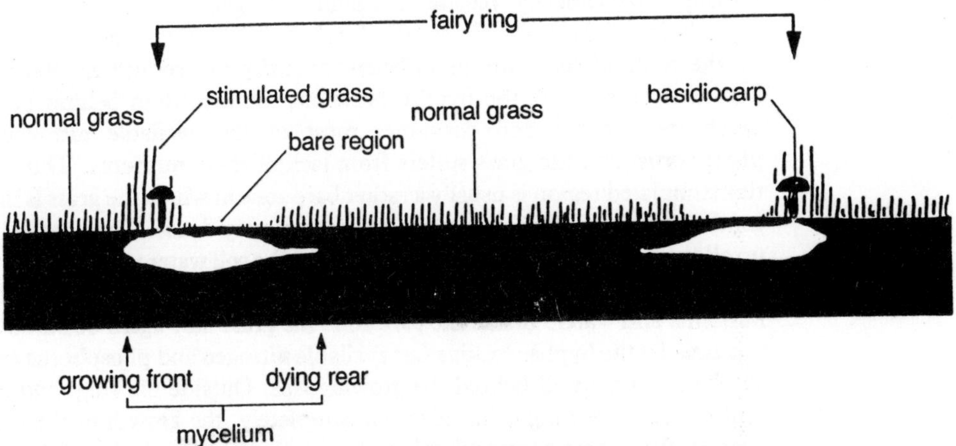

Fig. 9.3. Diagram of fairy-ring as seen in sectional view: normal soil shown black; soil occupied by the fungal mycelium shown white.

Wood ash has a very high pH and many of these fungi are known to have high pH requirements.

We may now consider fungi in pastures and on lawns. The species tend to be distinctive and few of the larger species are common to both grassland and woods. Especially mycorrhizal types are absent. Particularly characteristic are species of the large genus *Hygrocybe*. These are agarics of medium size. A few are white, but most are brightly coloured red, yellow, orange or even green, and, being slimy, tend to shine brightly in good light. *Agaricus campestris*, the field mushroom, is common in pastures as well as *A. arvensis*, the horse mushroom, and in acid pastures and grasslands *Lepiota procera*, the parasol mushroom. These last two are amongst the largest agarics known, the cap being up to 250 mm in diameter. They all live on decaying roots of grasses and aboveground parts incorporated into the soil. The two former fungi are particularly abundant where cattle and horses have grazed. Another abundant species is *Marasmius oreades*, the fairy-ring fungus.

Fairy-rings are the circles of dark green grass so frequent on lawns, golf courses and downs. They range from less than a metre up to 200 m in diameter. In the late summer basidiocarps occur on the ring. Although *M. oreades* is the commonest species involved, many other agarics, including *A. arvensis* and *L. procera*, can form rings, as can puff-balls occasionally.

In fairy-ring formation a mycelium becomes established in rather uniform soil. It then grows outwards in the upper layers of the soil forming a circular colony in much the same way as a mould produces a circular growth in a petri dish. Only the outer edge grows, the older parts, 200–300 mm further back, progressively dying (Fig. 9.3).

M. oreades is known to be able to degrade the humus fraction of the soil and to mineralize the organic nitrogen and phosphorus associated with this. In the immediate neighbourhood of the growing margin of the colony, the hyphae initially enzymatically release inorganic nitrogen and phosphorus from degradation of the humus in excess of their requirements. They are absorbed by the roots of the grass in sufficient quantity to produce an abnormally luxuriant growth. As the hyphae grow on, the mycelium behind becomes much more dense and consumes most of the available nitrogen and phosphorus, and the grass suffers from lack of these nutrients. Thus inside this stimulated region is usually a rather bare zone in which the grass is in poor condition or even dead, where the hyphae are very dense. They not only use up all the available water but appear to make the soil water repellent such that it does not easily wet up. Thus the grass suffers from shortage of both mineral nutrients and water. Inside the bare zone the grass may again be stimulated, this time by the hyphae leaking out available nitrogen and phosphorus as they autolyse and die well behind the growing tips. Outside the ring, and in the centre where the fungus has died out completely, the growth of the grass is normal. When basidiocarps develop, they are found to the inside of the ring of lush grass. The ring enlarges at a radial rate of 100–150 mm a year. On this basis a few of the largest rings must be several centuries old. Fairy-rings are not limited to grassland. They also occur occasionally in woods, but then are visible only when basidiocarps develop in autumn.

In pastures toadstools are often associated with dung. Further there is a well-known flora that develops on the dung of such herbivores as horses, sheep and rabbits. If freshly deposited dung is kept reasonably moist under a transparent cover with provision for aeration, a rich fungal flora develops showing a fairly regular succession. This succession is easily observed and the fungi involved are especially attractive. It has been studied intensively by mycologists because it provides the opportunity to investigate experimentally the interactions of a considerable variety of fungi in a natural, but strictly limited, habitat.

In this succession sporangiophores of Mucorales, notably *Pilaira*, *Pilobolus* and *Mucor*, appear in the first two or three days. They are soon accompanied by small apothecia of several Discomycetes, especially species of *Ascobolus*; and by this time the sporangiophores of the mucoraceous moulds are often parasitized by other moulds of the same order, especially *Chaetocladium* and *Piptocephalis*. Next to appear are tiny perithecia of *Podospora* and *Sordaria*. After ten days or so small agarics, notably species of *Coprinus*, *Stropharia* and *Panaeolus*, begin to dominate the dung. Some Deuteromycotina also commonly occur, including the eelworm-trapping *Arthrobotrys oligospora*. For most of the coprophilous fungi the dispersal story is the same as already described for *Pilobolus* (p. 30). The stages (mucoraceous, discomycete, pyrenomycete and hymenomycete) in the succession appear to be determined by the time required from germination of the spores to the production of reproductive structures. However, it is not quite clear what are the factors responsible for the end of a stage. Nevertheless there is good evidence that in the final or hymenomycete stage, the dominance of species of *Coprinus* may be due, in part at least, to their mycelia having an antagonistic effect on the hyphae of other fungi. When hyphae of *Coprinus* make contact with the hyphae of other fungi in the dung, the contacted cell and the immediately adjacent ones undergo vacuolation and loss of turgor. There is also a drastic increase in the permeability of the cell membrane. These lead to the death of the cell. This phenomenon is known as hyphal interference and is a very effective form of antagonism. It must be emphasized that the whole system is complex. In addition to a large fungal population, there is a considerable bacterial population as well as a fauna of small animals such as eelworms and insects.

Although the coprophilous succession is fairly well defined in the laboratory, it may not follow such a regular course in the field. Further, even in the laboratory, it may be greatly modified by conditions such as the prevailing relative humidity.

Study of the distribution of the larger fungi in various habitats relies largely on recording the occurrence of the reproductive structures, since the mycelia are concealed in the substratum. In such studies it is necessary to recognize that there is a phenology, the study of the time of reproduction, of these organisms. In general terms the maximum development of the larger fleshy fungi is in the late autumn, although there are some, such as the morel (*Morchella*) and St George's mushroom (*Tricholoma gambosum*) which appear in late spring. The first severe frosts of late autumn usually put an end to the main toadstool season. There is, however, one outstanding winter species, namely *Flammulina velutipes* (p. 89). In this agaric the basidiocarps, even in a fully hydrated condition, can survive being frozen stiff. On thawing spore liberation soon begins again. This species is common throughout the severest winters.

There have been few quantitative studies in fungal phenology. One such study concentrated on the agarics in the grass sward of an apple orchard during a period of 5 years. For each 10-day period, the numbers of basidiocarps of individual species were recorded. The commonest, all typical grassland species, were: *Panaeolus foenisecii*, *Conocybe lactea*, *C. tenera* and *Coprinus plicatilis*. *Agaricus campestris* also occurred. The mean percentage occurrence of each of these species over the 5-year period is shown in Fig. 9.4. They varied in time of reproduction from early summer to autumn.

The larger fungi of woodland and grassland have their mycelia in the soil accompanied by many microfungi which seem to contribute the major element of the fungal community of soil. This has been extensively studied throughout the world and seems to be dominated by species of *Mucor*, *Penicillium*, *Aspergillus* and *Trichoderma*. In the study of the microbial population of soil long lists of fungi found there have been compiled. In this field, however, the results obtained must always be considered having regard to the technique employed.

The poured-plate procedure was much used in earlier studies. In this a small amount of soil is suspended in sterile water, the coarse material being allowed to settle. Then 1 ml of the suspension, suitably diluted, is mixed with 10 ml of sterile liquid agar cooled to 40°C before being poured into a sterile petri dish where the agar sets to a jelly. The fungal colonies developing can then be counted, isolated and identified. Here the results depend on the nature of the agar used, and on the ability of the fungi present to grow relatively

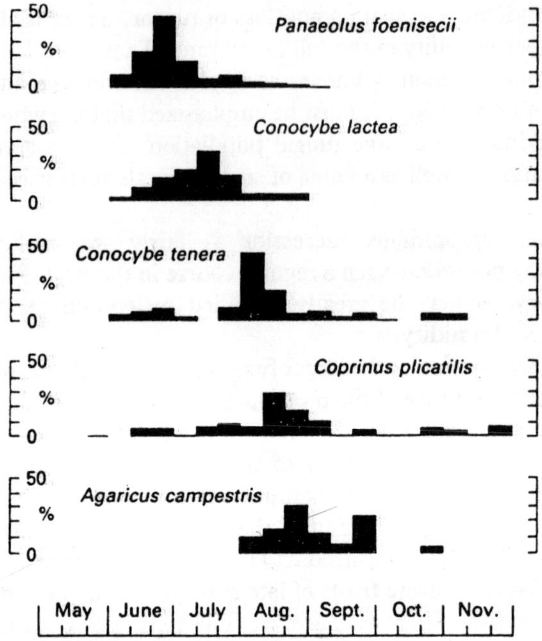

Fig. 9.4. Seasonal occurrence of toadstools in orchard sward. Histograms show percentage of total number of basidiocarps of each species in each 10-day period averaged over years 1964–68. (After Bond 1972)

quickly so that they figure in the record. Another serious objection is that fungi associated with the larger soil particles, which settle when the soil suspension is being prepared, tend not to be recorded.

In more modern work the soil-plate method is favoured, although it is still open to many of the objections to the earlier procedure. A sample (5–15 mg) of soil is taken up on a flamed, flattened inoculating wire. This soil is crushed and spread over the bottom of a sterile petri dish, using, if necessary, a drop or two of sterile water to assist dispersion. About 10 ml of sterile agar at 40°C is added, and the soil further dispersed by the wire before setting occurs. With this method many of the fungal colonies develop from soil particles.

Other methods have been used and always the species isolated depend to a great extent on the procedure. Again the isolation of a fungus on agar gives no information about how it occurs in the soil, whether it is there as dormant spores or as an active mycelium.

It is extremely difficult to determine the state in which fungi exist in the soil. A cumbersome but elegant method involves impregnating a block of soil with a hard-setting resin. From this as from a rock, thin sections can be produced by petrological methods and the spatial relationships of the fungi and the soil particles are then directly visible. However, although mycelium, spores and sometimes even whole spore-bearing structures can be seen, identification is rarely possible. Again this static picture of the fungi gives no information about their activity.

A knowledge of the different kinds of fungi that can be isolated from soil may be of value, especially if their nutritional requirements are known. Soil is a dynamic system to which new organic matter is continually being added, mainly as vegetable matter. In this the simpler carbohydrates, such as sugars and starch, are rapidly utilized by certain fungi which have spores that germinate readily and mycelia that grow rapidly. These are the 'sugar fungi', in the main members of Mucorales, most of which cannot hydrolyse cellulose. They are followed by other species belonging to Ascomycotina and Basidiomycotina capable of dealing with cellulose. A few of these can also break down lignin. The most refractory organic matter remains, at least for a time, as amorphous humus. Although it had been obvious to mycologists for a long time that the mycelium of toadstools must be common in many soils, it was only with the introduction of the soil-plate method that Basidiomycotina began to appear in lists of fungi isolated from soil.

Within the soil, fungi compete with one another and with other soil organisms. Thus *Trichoderma viride*, an abundant species in soil, excretes specific substances capable of killing the mycelia of other fungi. Competition of fungi with one another in the soil may be of importance in plant pathology. In this connection special mention may be made of *Gaeumannomyces graminis*, responsible for take-all of wheat, and *Fusarium oxysporum* which causes wilting in a wide variety of crop plants. These fungi can exist as saprotrophs in the soil, but are there subject to competition with more vigorous species which are purely saprotrophic. The relationships between these types, which can be affected by soil treatment, is of considerable practical significance.

Within the soil special conditions exist on and near the surface of living roots. Mycelia of various species often occur on the root surface taking advantage of organic substances diffusing from the surface cells. Such saprotrophic existence may be a preliminary to parasitic attack. This particular environment is the rhizoplane. There is also an enriched zone close to the root where conditions are different from those in the bulk of the soil. This is the rhizosphere which tends to have a rich fungal community.

Temperature is one of the cardinal factors which determines the distribution of many fungi. Compost heaps, damp hay ricks, piles of stable manure and industrial wood chip piles are all self-heating systems which encourage a distinctive fungal population as the temperature rises above 30°C. Most of the fungi involved cannot develop below 20°C, but grow vigorously at 35–45°C. They can be isolated by adding tiny pieces of debris from compost to dishes of nutrient agar which are then incubated at 45°C, taking care to prevent desiccation. Outstanding examples of this small assemblage of fungi are *Mucor pusillus*, *Chaetomium thermophile* and *Thermomyces lanuginosas*. The latter can grow at 60°C. Even alligator nests which are also self-heating support such fungi. Apart from their temperature requirements, they differ little from many other fungi. Some, such as *C. thermophile*, are markedly cellulolytic and are important decomposers in such systems. Some, because of their ability to grow at 37°C, have been reported as causing diseases of warm-blooded animals. *M. pusillus* has been associated with mastitis in cattle and with a variety of mycoses.

Fungi are predominantly terrestrial, less than 2% of the species being aquatic. Most of these reproduce by zoospores, and it is reasonable to regard them as primitive aquatics whose ancestors have always led an aquatic life. Amongst these fungi, species of *Saprolegnia* and *Achlya* occur on floating or submerged organic matter, especially if it is rich in protein. In ponds and canals a floating fish with a white fringe of saprolegniaceous mould is a familiar sight. In highly polluted waters *Leptomitus lacteus*, the sewage fungus, is abundant and sometimes blocks sewage filters. It is peculiar in fungi that it cannot utilize simple sugars, such as glucose, and requires organic sources of nitrogen. It can utilize both alanine and leucine as sole carbon and nitrogen sources but more suitable carbon sources include acetate, pyruvate and fatty acids which are all readily available in sewage. Again microscopic chytrids are widespread not only as parasites of algae, including especially the unicells of the phytoplankton, but also as saprotrophs on dead submerged organic materials consisting of cellulose, chitin and keratin. A number of interesting fungi develop as white pustules on fruit with a smooth surface (e.g. apples, rose-hips, tomatoes) if these are kept submerged in a pond for a few weeks. The commonest fungus obtained in this way is *Blastocladia*.

It is to be noted that many of the fungi usually regarded as aquatic can be isolated from soil; a real meeting-ground of aquatic and terrestrial organisms.

As well as the primitive aquatic fungi of freshwater habitats, there are others that have probably been derived from terrestrial ancestors. The abundant Hyphomycetes regularly to be found on submerged decaying leaves of trees and shrubs in streams and rivers have already been mentioned (p. 114). These fungi clearly play a major role in the ecology of well-aerated waters in which the leaves of deciduous trees provide the major annual intake of organic matter. Soluble simple organic carbon compounds are readily leached from these leaves when they become immersed in streams. They would be expected to use polymers such as cellulose. Many have been shown to do so. The majority also produce pectinases and hemicellulases and the production of such enzymes may be one of the main reasons why fungal activity is most significant in the early stages of decomposition, soon after the leaves fall in. As in terrestrial ecosystems only a small fraction of the nutrients locked up in leaf material is available to be directly exploited by the micro-fauna within the ecosystem. Usually more than 60%, and often as much as 80–90%, of the leaf-litter eaten by such detritus feeders is returned in the form of faecal pellets. To enable them to gain access to the remainder of the leaves the micro-fauna in streams rely upon the aquatic Hyphomycetes to degrade cellulose and other polymers and utilize these to increase their own biomass. Fungal carbohydrates, such as trehalose and glycogen, fats and proteins are then digested by the micro-fauna. Fungal mycelium is also rich in choline, B vitamins and ergosterol which are all beneficial to the micro-fauna. The fungi may also concentrate biologically important minerals. Many aquatic animals thus depend upon the increase in fungal biomass as nutrient sources. More so it has been shown that the detritus feeding micro-fauna in streams prefer leaves partially decomposed by the aquatic Hyphomycetes to sterile or freshly fallen ones. Thus they have a decisive influence on food selection. The aquatic

Hyphomycetes are not only the major decomposers of deciduous tree leaves in freshwater streams but they are also very important intermediaries in the food chain.

There is also a rich array of saprotrophic freshwater Ascomycotina, especially on the dead submerged stalks of reed-swamp plants and on water-logged sticks in lakes, but there are very few aquatic Basidiomycotina. The ascus as an explosive sporangium is sufficiently powerful to function underwater whereas the basidium, although exposed, is covered to protect it from rain as it is unable to function if wetted or submerged.

Some fungi are marine, but their numbers are relatively small and they represent a tiny element in the total of living organisms in the sea. A few are chytrids and similar microscopic fungi. However, the great majority of the fungi known from the sea are Ascomycotina, particularly Pyrenomycetes. Most of these are wood-destroying fungi that have been collected by submerging small blocks of wood in the sea for considerable periods.

Fungi as plant pathogens

10

Plants are subject to disease. Some diseases result from disturbed metabolism produced, for example, by a deficiency of essential mineral elements in the soil. In most however a virus or a bacterium or a fungus is the pathogen. Fungi, rather than bacteria, are the most widespread and destructive parasites of plants. This is the reverse of the situation in man. For him, bacterial diseases are much more important and serious.

Fungal plant diseases are more prevalent and certainly more spectacular amongst cultivated crop plants. These are grown in stands of a single cultivar or variety of a species. They are all genetically identical so if one is susceptible, all will be.

In adopting parasitism as a mode of life, fungi derive their organic requirements from living plants or animals. They come into association with their host and share a common life. A symbiosis, a living together, is established but only the fungus benefits. The host is harmed. Such a relationship is called a parasitic symbiosis.

Parasitic fungi can be classed in a number of ways. Some are very specialized and can only grow in association with a living host. They must be associated with living cells of the host in order to grow and reproduce. They are termed obligate parasites. In others this association is not absolutely necessary. They are less specialized and able to grow away from their hosts as competitive saprotrophs. They are termed facultative parasites. Parasitic fungi can perhaps be best classified by their modes of nutrition. Two extreme modes can be recognized – nectrotrophs and biotrophs.

Nectrotrophic fungi derive their organic nutrients from dead cells which they have killed. They are 'dead feeders'. They are destructive parasites, differing from saprotrophs only in that they kill host cells. They often kill their host quickly. Many necrotrophs are facultative parasites which can live as competitive saprotrophs on the dead organic matter in their environment. The mode of nutrition of such fungi depends very much on the conditions in which they find themselves. By contrast, biotrophs are 'living feeders'. They are only capable of deriving their nutrition from living host cells. The fungus either dies or goes into the survival state should the host, or that part of it infected by the fungus, die. Away from their hosts, biotrophs are only found as dispersal or dormant spores. A very balanced and long-lasting relationship may be established with the host. Necrotrophy and biotrophy are two nutritional extremes. Some parasites, such as *Phytophthora infestans*, which causes potato blight, are truly intermediate, being at first biotrophic and then necrotrophic.

A very large number of fungi from a diversity of groups are necrotrophic plant parasites. The less specialized of these, in addition to being aggressive parasites, have a prolonged and very successful saprotrophic phase on other dead substrates. With increasing specialization towards parasitism, there is a tendency for the progressive loss of this competitive saprotrophic ability. Most grow in pure culture and have simple nutritional requirements. This lack of any specialized nutritional requirements means that the majority of necrotrophic fungi are capable of attacking a wide range of hosts. This is aided by the fact that host tissues are killed in advance of the hyphae such that any actively induced host resistance mechanisms may not be provoked. The majority cannot overcome the resistance mechanisms of mature plants. Often the only kinds of hosts which can be attacked are seedlings, wounded or stressed plants or ripe fruits and storage organs – immature, sick and senescent tissues. A large inoculum level is required to make an effective attack.

Necrotrophs kill host cells before they penetrate them by the secretion of extracellular cell-wall degrading enzymes or toxins or both. The effects of these are responsible for the disease symptoms seen. Most necrotrophs also appear to be insensitive to any toxic metabolites produced by the death of host cells. Their effects vary from local discrete lesions, as seen in some leaf spot diseases, to massive tissue destruction, as seen in fruit rots.

Many cause disease in which the major symptom is a soft watery rot of parenchymatous tissue. This is seen in such diseases as damping off in seedlings caused by *Pythium* spp. and fruit rots caused by *Monilinia fructicola* and *Penicillium expansum*. The major substrates attacked by these parasites are the pectic substances in the middle lamella between cells. On infection the tips of the advancing hyphae produce copious amounts of a variety of pectolytic enzymes. This leads to cell separation and partial degradation of the host's cell-walls but more important for the fungus the release of the galacturonic acid building blocks of the wall. The galacturonic acid residues are the main source of organic carbon used by these fungi for their growth and reproduction. The pectolytic enzymes have another very important effect. They bring about a marked increase in the permeability of the protoplasts of the host. The cells leak so that they lose turgor and rapidly die. The fungus then also uses the organic solutes which pour out of the cells as a further nutrient source. Both the separation of the cells and the release of the watery solution from the cell vacuoles are the main causes of the symptoms seen, a soft watery rot, but the release of the vacuolar solution aids the diffusion of the pectolytic enzymes in advance of the growing hyphal front.

Many of these diseases can easily be demonstrated. If any seeds, such as those of cress, lettuce or cabbage, are sown densely in a pot of garden soil and are kept thoroughly damp, when the seedlings emerge some are soon weakened at soil level or just above, following attack by *Pythium* spp. The cortical parenchymatous cells of the hypocotyl are separated, lose turgor and become water-soaked. The seedlings topple over and very rapidly rot. Some seedlings never emerge. They are rotted before they do so. *Pythium* is an Oomycete which has a prolonged saprotrophic phase in the soil where it produces motile zoospores, resting oospores and chlamydospores on hyphae growing through the soil.

Root exudates, such as simple sugars and amino-acids, are important in infection. They stimulate oospores and chlamydospores to germinate and induce chemotropic growth of hyphae and also chemotactic movement of zoospores towards the root. The uptake of these exudates also increases the capacity to infect by providing nutrients for growth of hyphae and so building up an inoculum. The severity of disease increases with increase in soil moisture levels up to saturation. High moisture levels stimulate zoospore production and free water is required for their dispersal. They also restrict gaseous exchange. Whereas the host is stressed by low oxygen levels, *Pythium* is more tolerant of low oxygen and high carbon dioxide tensions than most soil fungi. It then has an advantage over its competitors and is more vigorous. Water-logged roots also produce more exudates which again favour attack by *Pythium* and, being stressed, are more susceptible. Thus the vigour of both the host and the fungus are determined by environmental factors. Under drier regimes, the seedlings are more vigorous and produce less exudates and, as a consequence, are less susceptible to attack. This example illustrates the close interactions which occur between host, parasite and the environment in disease expression.

On arrival at the root surface, the zoospores encyst and soon germinate to produce hyphae. These hyphae, the vegetative hyphae attracted to grow towards the roots and those from the germinating resting spore grow randomly over the root surface. If the seedlings are both overcrowded and water-logged, the hyphae grow up the root and over the surface of the soil, without being subjected to desiccation, and form a sparse network over the surface of the hypocotyl. At this stage a sufficient inoculum has been built up to attack the host. Hyphae penetrate at a number of points and spread between the cortical cells as they become separated as a result of the copious production of a variety of pectolytic enzymes by the fungus. The cells leak, become water-soaked and lose turgor. As turgor is the main means of support the seedlings collapse and rapidly die.

Many fungi affect ripe fruit. Raspberries, strawberries (Plate 3a) and blackberries are commonly attacked in damp weather by grey mould, *Botrytis cinerea*, as are courgettes, marrows and cucumbers. Apples, pears and plums are at the mercy of a whole range of soft-rot fungi, especially *Rhizopus stolonifer*, *Monilinia fructicola* and *Penicillium expansum*. Apples and plums rotted by *M. fructicola* are abundant in orchards in the summer and early autumn both on the tree and on the ground. Unlike *Pythium*, the majority of these soft-rot fungi can only enter through a wound, such as that caused by a bird peck or hailstone. Following infection by *M. fructicola*, a brown rot rapidly spreads through the flesh of the fruit and soon easily recognizable concentric patterns of conidial cushions appear on the surface (Plate 3c and d). Under the microscope this stage is easily identified by the branched chains of colourless conidia (Fig. 10.1). The symptoms produced by *M. fructicola* and *P. expansum* are quite different although they may produce similar pectolytic enzymes. *M. fructicola* causes a firmer, darker rot which is faster spreading than the rot caused by *P. expansum*. The rot is faster spreading in *M. fructicola* because the fungus grows faster. In such necrotrophs, cell damage on infection releases phenolics which are oxidized by host phenolase enzymes to give brown products. These latter form complexes with the pectolytic enzymes which are inhibited and cell separation is thus limited as is leakage from the cells. The fungus grows on but its pectolytic enzymes are inhibited virtually as soon as they are produced. The result is a firm, brown rot. *P. expansum* produces an inhibitor of the phenolases so there are far fewer brown products and the pectolytic enzymes bring about complete separation of the cells and leakage from them is much greater to give a pale, soft and watery rot.

A large number of other enzymes capable of degrading other components of the cell-wall can be isolated from culture filtrates of these fungi and rotted tissues. These include cellulase and hemicellulases. They are probably of secondary importance to most rot-causing fungi in terms of disease expression but they may be very important at a later stage to enable the fungi concerned to live as competitive saprotrophs.

Trees are often killed by necrotrophic parasites which then live on the trees as decomposer saprotrophs for a number of years. A number of Polyporaceae or bracket polypores, such as *Piptoporus betulinus* and *Ganoderma adspersum* have already been mentioned as wound parasites. *P. betulinus* is specific to birch (*Betula*). Windborne basidiospores impact onto the xylem exposed at wound sites where branches have broken in the wind. In the wood, cellulase and hemicellulases are produced as the fungus grows down into the trunk and a typical brown rot develops (Chapter 9). With the structural cellulose microfibrils removed, the tensile strength is lost and the trunk snaps off under high winds at about 3 m above ground. The fungus continues to decay the trunk as a saprotroph, producing its basidiocarps on the fallen and standing parts of the trunk for a number of years. Other wound parasites, such as *Polyporus squamosus* on sycamore (*Acer*), cause a white rot. They attack the heartwood, rotting it with the result that major branches break off and the whole tree may topple in the wind.

In such instances the distinction between necrotrophy and saprotrophy is blurred as the death of the tree is a consequence of the destruction of the dead, water-conducting xylem. Other Basidiomycotina can colonize and kill healthy, undamaged trees. Clumps of basidiocarps of *Armillaria mellea*, the honey fungus, are common on and around dead tree stumps in the autumn (Plate 2b). The stump or the dead trunk and roots are the operational saprotrophic base for the parasite to act. Black, water-proof, root-like rhizomorphs grow out, 10 m or more, from the base until they make contact with the roots or trunk base of another living tree (Fig. 10.2). These rhizomorphs are aggregates of several thousand hyphae. Specialized hyphae within conduct nutrients from the colonized food base to the new victim.

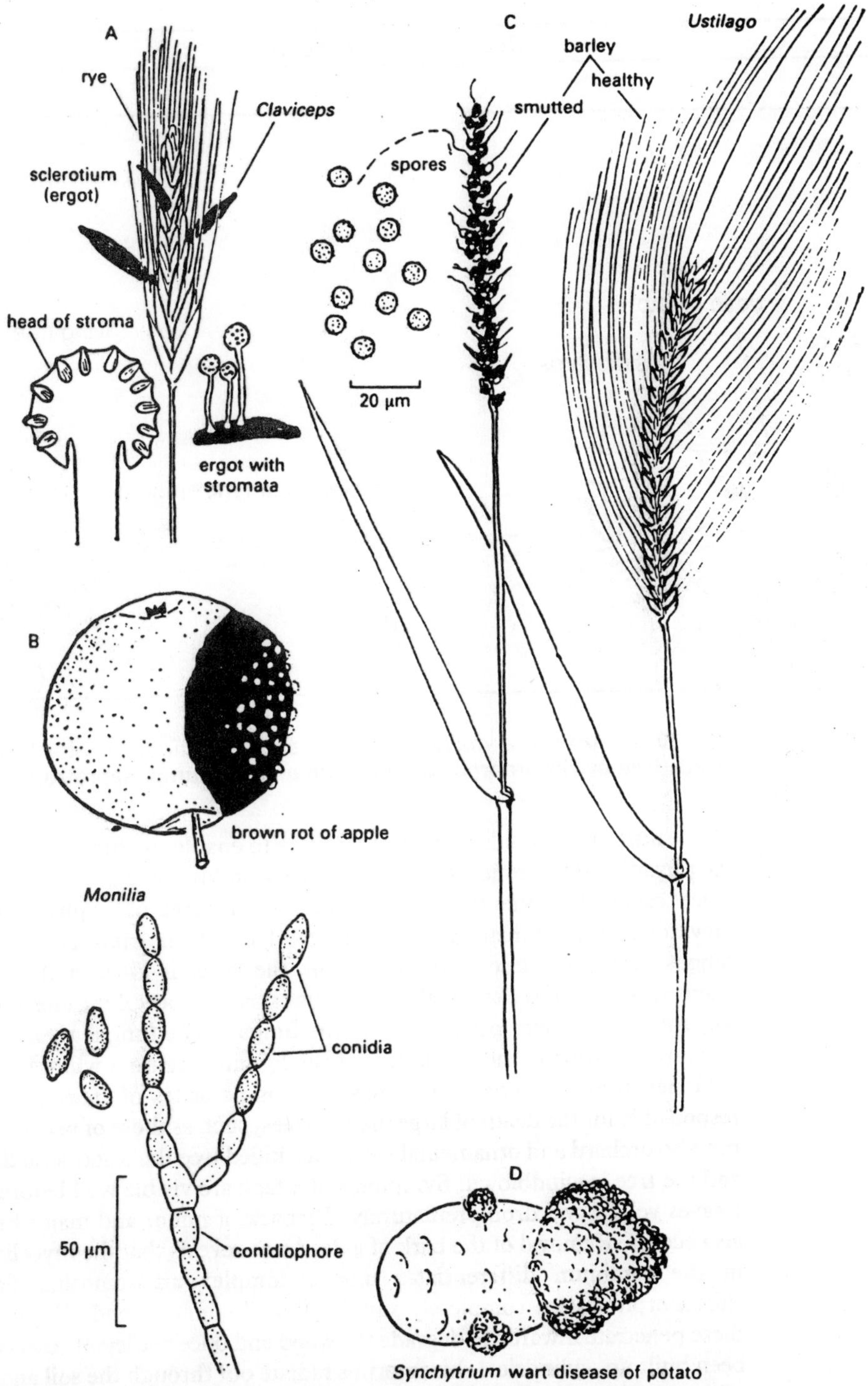

Fig. 10.1. Plant diseases. A, ergot of rye (*Claviceps purpurea*): three sclerotia (ergots) on rye; ergot producing three stalked perithecial stromata; and one stroma in longitudinal section showing the perithecia. B, brown rot of apple (*Monilina fructicola*); partially rotted apple with cushions of conidia on the dead region; and the conidial apparatus magnified. C, loose-smut of barley (*Ustilago nuda*): healthy and smutted heads and some teliospores. D, wart disease of potato (*Synchytrium endobioticum*).

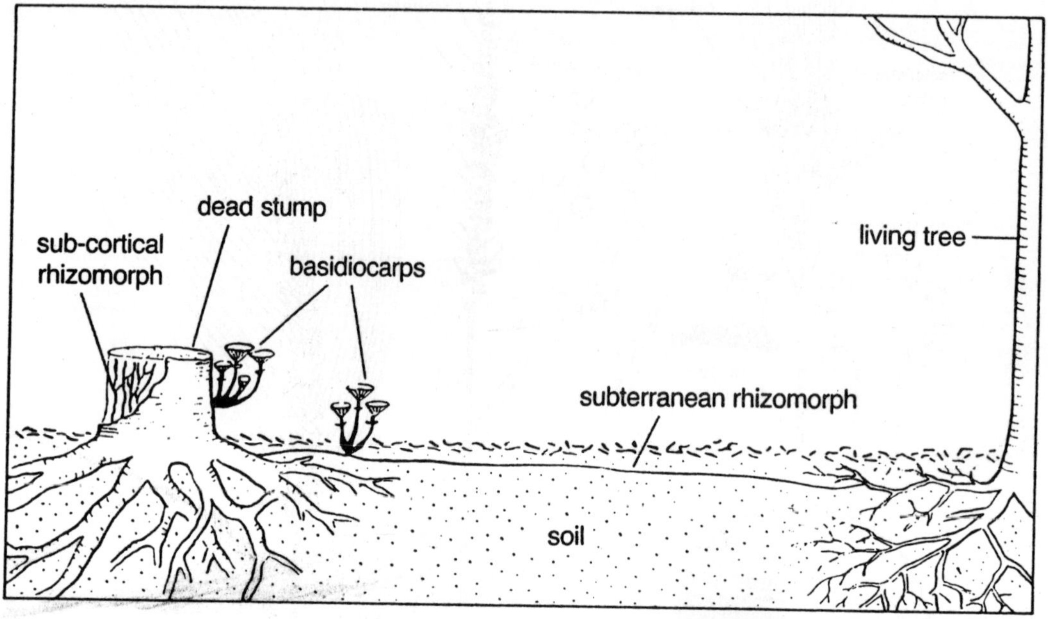

Fig. 10.2. *Armillaria mellea.* Diagram showing mode of attack on living beech tree by rhizomorph coming from a dead stump (saprotrophic base).

These nutrients provide a massive resource to enable the fungus to overcome the physical and chemical barrier of the bark and become established within. Once established the mycelium invades as a sheet of hyphae spreading upwards in the cambial region. When the base of the trunk is reached, the fungus girdles the tree in the cambium, the latter is killed and the tree is doomed. Death of the cambial cells can be caused by pectolytic enzymes, as in soft rots, or by toxin production by the fungus. The fungus then lives as a saprotroph utilizing the cellulose and the lignin to cause a white rot.

Honey fungus is one of the most serious parasites of trees often being responsible for the death of large mature trees, such as those of beech (*Fagus*), but also orchard and ornamental trees. In a killed tree the wood soon dries out and the tree is windblown. Symptoms of attack are visible well before death. Leaves yellow and drop prematurely. Dieback of minor and major branches also occurs. Removal of the bark of a dead tree reveals that the mycelial sheet in the cambium differentiates into a complex anastomosing flattened rhizomorph system completely surrounding the inner wood. Hyphae from these penetrate inwards to degrade the wood and once sufficient reserves have been built up, cylindrical rhizomorphs radiate out through the soil and attack other trees. The food reserves of a single spore are inadequate to establish the fungus on an intact root whereas the rhizomorph has a huge inoculum potential which can be brought to bear effectively on a healthy root.

Heterobasidion annosum shows a different strategy. It can be devastating once established in a young pine plantation, killing young trees up to 25–30 years old outright and causing a fibrous, red–brown butt- or heart-rot spreading up from the base of more mature trees. Basidiocarps, hard, woody brackets, occur at the base of killed trees. It becomes established on freshly cut stumps, such as occur during thinning operations in young pine plantations. Such stumps are particularly selective to basidiospores of *H. annosum*. The spores establish and grow down into the dead roots building up a sufficient inoculum to attack others. When a dead, infected root makes direct contact with a living healthy root, infection occurs and the fungus spreads up the tree girdling the stem base in the cambium and killing it in young trees but remains in the wood and causes a heart-rot in older trees. Here again a saprotrophic base is needed for parasitic attack.

Cell death is an obvious symptom in soft rots of parenchymatous tissues. Many necrotrophic pathogens can also bring about cell death by the production of pathotoxins. Fungi produce a wide range of toxic substances in culture. They are usually secondary metabolites and may kill plant cells. However, most of these are not produced *in vivo* on infection and play no role in the expression of disease symptoms. Pathotoxins, by definition, must have a role in disease. The idea that pathogenesis might be due to the production of toxins is by no means new. The idea first developed by analogy with medicine in the nineteenth century at the time when Pasteur showed that the symptoms of diphtheria and tetanus were largely due to toxins produced by the pathogen.

Pathotoxins are produced in minute amounts and are mobile within the plant such that disease symptoms can be produced at some distance from the infection site. They may cause small discrete necrotic lesions on leaves, massive chlorosis or kill the whole plant. Chemically they are very varied. They may be polysaccharides, organic or fatty acids, peptides or glycoproteins. They may be the sole determinants of disease or may be part of a series of complex interacting factors causing disease.

Alternaria solani causes the *Alternaria* blight of potatoes and tomatoes. Small circular black lesions occur between the veins of the leaves. Heavily infected leaves show rapid chlorosis of uninfected tissues between the spots, followed by early death of the leaves and the shoots. The fungus produces a low molecular mass dibasic acid, alternaric acid, which is responsible for the disease. Other species of *Alternaria*, such as *A. tenuis*, also cause chlorosis, followed by death, of many seedlings, by inhibiting chlorophyll accumulation. A cyclic tetrapeptide, called tentoxin, is produced by the fungus which acts by blocking ATP synthesis preventing the formation of essential proteins involved in the development of chloroplasts. Toxins such as alternaric acid are non-specific in that they will affect a wide variety of plants which are not parasitized by the fungus. Others are remarkably host-specific showing the same specificity as the parasite itself. *Helminthosporium victoriae* causes a devastating seedling blight of oats in the USA. It is particularly virulent on the cultivar called Victoria where it causes a wilt, with reddish necrotic lesions on the leaves and a rot at the base of the stem. It produces a toxin, named victorin. Susceptible cultivars, such as Victoria, are very sensitive to it but resistant cultivars are not affected at all. In this particular case not only does the toxin show the same specificity as the fungus itself but only toxin-producing strains of the fungus cause disease symptoms. It is the major determinant of disease.

Most of our elms have been killed by Dutch elm disease. The fungus responsible, *Ceratocystis* (*Ophiostoma*) *ulmi*, is not a wood-rot fungus but one of a group of fungi known as the vascular wilt fungi. They cause many devastating and spectacular diseases by bringing about the blocking of the water-conducting elements of the xylem. Fungi in the genera *Fusarium* and *Verticillium* are the best known of the soil-borne wilt-causing fungi infecting a very wide range of plants world-wide, including peas, tomatoes, hops, bananas and cotton. They are all wound pathogens. *C. ulmi* is spread by elm bark beetles feeding in the canopy. Inside the plant the fungus becomes restricted to the xylem but toxins are certainly not the sole determinants of disease. In infected plants the actual resistance to water flow through the xylem is very substantially increased. Flow may be less than 5% of that in healthy plants. Some of the resistance can be explained by physical blocking of the xylem vessels, in part by the fungus and in part a response of the host to infection. Hyphae may amass in the vessels and also produce mucilaginous polysaccharides which achieve some blockage. The fungus may also produce some cell-wall degrading enzymes, weakening the walls of some of the vessels and causing them to collapse under tension. The host also makes matters worse by itself producing gums and mucilages and by forming tyloses. The protoplasts of the adjacent parenchyma balloon out into the lumen of the xylem vessels, restricting water transport and delaying spread of the fungus. Nevertheless all these fungi produce toxins, such as fusaric acid and others by species of *Fusarium* and cerato-ulmin by *C. ulmi*. Fusaric acid, like many other toxins, increases cell membrane permeability, giving rise to leakage of vital ions. One of the best known wilt-inducing toxins is fusicoccin, produced by *Fusicoccum amygdali* involved in the wilt of almonds and peaches in the Mediterranean. Fusicoccin is a terpenoid glycoside which is transported in the transpiration stream. It causes irreversible opening of the stomata in the light and the dark. It affects ion transport across the guard cell membranes, stimulating uptake of potassium ions and the release of hydrogen ions. As a consequence, the guard cells become more turgid than normal and remain so such that the stomata open wider and stay open. The plants become water stressed and in the dry Mediterranean eventually the leaves desiccate. Southern Corn Leaf Blight, one of the worst epiphytotics of modern times, is caused by a fungal toxin, T-toxin. Early this century all cultivars of maize were self-fertile, but by the early 1960s, a new gene had been introduced into maize, the Texas cytoplasmic gene, for male sterility (Tms gene) into virtually all cultivars to simplify the production of hybrid varieties. Until then, *Helminthosporium maydis* had been known as a weak necrotrophic pathogen causing only little damage but it mutated to form a new race, race T. This produced massive necrotic lesions on all cultivars carrying the Tms gene. Virtually the whole maize crop in America was wiped out in 1970. Total crops browned off and appeared as if they had been scorched by fire. The Tms gene is a mitochondrial gene and the T-toxin specifically affects the mitochondria of cultivars carrying this gene. Mitochondria of normal maize are not affected. It increases the permeability of mitochondrial membranes and stimulates oxidative reactions in the mitochondria. Electron transport is uncoupled from ATP synthesis depriving the plant of energy needed for biosyntheses.

Biotrophic plant parasites occur mainly in three orders of fungi from three separate divisions of the fungi. These are the Peronosporales, the downy mildews, in the Mastigomycotina, the Erysiphales, the powdery mildews, in the Ascomycotina and the Uredinales, the rusts, in the Basidiomycotina. Of these, only a few of the rusts have as yet been cultured in the absence of living host cells. Such obligate parasites as these are nutritionally very exacting and cannot derive nutrients from dead cells. They completely lack the ability to grow and compete as saprotrophs.

These fungi arrive on the surface of the leaves or shoots of susceptible hosts as spores. Once arrived, free water is normally necessary for the spores to germinate, although the conidia of some powdery mildews provide a conspicuous exception to the rule. Droplets of water on the epidermis of the host resulting from dew or rain provide potential infection drops. If infection is to occur these must persist long enough to allow penetration to take place. Apart from their duration, other factors may operate in the infection drops for substances may diffuse into them from the living cells of the host, which stimulate or retard the germination of the spores. Penetration can occur by two distinct methods in leaves and soft shoots. They may penetrate either directly through the cuticle and cell-wall or via natural openings such as stomatal pores. The germ tubes of the conidia of powdery mildews, such as *Erysiphe graminis*, make contact with the cuticle and swell to form an ovoid, limpet-like appressorium (Fig. 10.3D). From the underside of the appressorium a very narrow infection peg penetrates the cuticle and cell-wall. Penetration is partly enzymatic and partly mechanical. In many pathogens a cutinase is liberated in response to the presence of cuticle. Basal levels in spores break down cuticle and release monomers on contact with cuticle. This stimulates the switching on of the fungal gene for cutinase production. On penetration and the utilization of the monomers, the gene is switched off again. Mutants lacking the gene are non-pathogenic. The formation of appressoria further illustrates the versatility of fungal hyphae in their ability to penetrate and permeate tissues. The firmly attached appressorium prevents any lift-off from the surface as the infection peg applies mechanical pressure, coupled with the production of lytic enzymes, as it bores through the cuticle and wall. Many necrotrophic pathogens and mycorrhizal fungi penetrate their host in a very similar manner.

The germ tubes of the urediniospores of many rusts, such as *Puccinia graminis*, and spores of downy mildews, such as *Bremia lactucae*, extend until they encounter a stomatal pore. Before actual entry an appressorium is formed over the pore (Fig. 10.3A, B).

After penetration, a swelling or vesicle usually arises from the tip of the infection peg and in the majority of biotrophs hyphal branches arise from this and grow between the host cells. They remain intercellular and do not penetrate the host cells. The cells are penetrated by very specialized branches of determinate growth and multiplicity of shapes called haustoria (Fig. 10.4

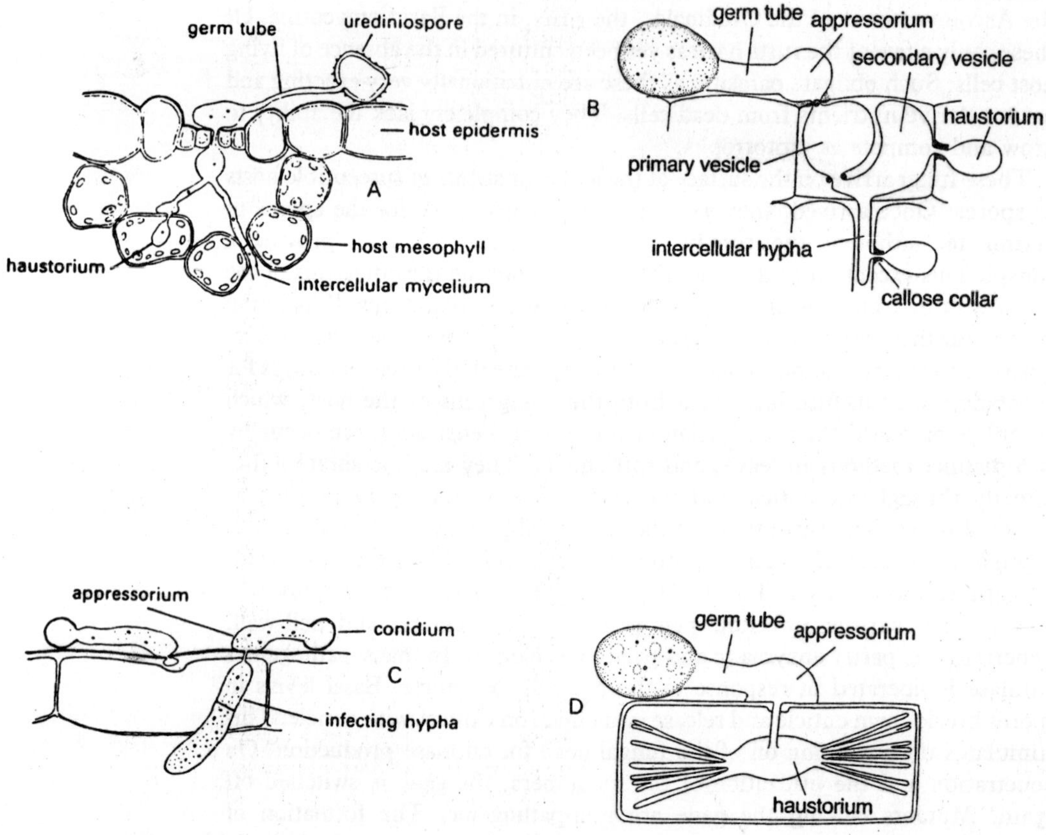

Fig. 10.3. A, germinating urediniospore of *Puccinia graminis* showing stomatal entry. B, germinating conidium of *Bremia lactucae* showing stomatal entry. C, germinating conidium of *Botrytis cinerea* showing direct entry. D, germinating conidium of *Erysiphe graminis* showing direct entry and a haustorium in the penetrated epidermal cell (all highly magnified).

and Plate 4d). These penetrate the wall but only invaginate the cell membrane. Haustoria are nucleated and rich in organelles such as mitochondria, ribosomes and vesicles, indicating that they are sites of high metabolic activity. They are presumed to be the structures responsible for the absorption of major nutrients from the host cells. In the Erysiphales, unlike in the other two orders of obligate biotrophs, all the hyphae are on the surface of

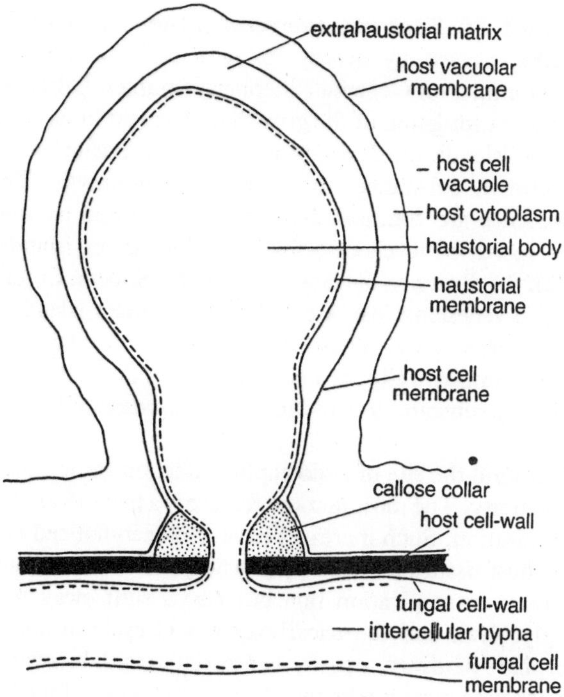

Fig. 10.4. Diagram of a haustorium of a biotroph in a host cell.

the host and only haustoria penetrate the cells, where they are restricted to the epidermal cells (Fig. 10.3D).

Biotrophic infections have a number of features in common. As the fungi concerned are dependent upon living host tissues to complete their development, they invariably cause minimal tissue damage, at least initially. They do not produce large quantities of extracellular cell-wall degrading enzymes or toxins. There is little or no immediate cell death. Cell-wall degrading enzymes are only produced when wall penetration occurs. Similarly toxins may be produced, although none have been recognized, in such small quantities that they increase membrane permeability, with some leakage of nutrients. Their success as parasites largely depends upon the extent to which they can extract the necessary nutrients for their growth and reproduction before their host is sufficiently damaged to decrease the synthesis of these nutrients.

Infection of a leaf or shoot creates a sink into which a variety of host-produced nutrients, especially sucrose, are rapidly removed. This induces the translocation of host metabolites into infected regions. These changed patterns of translocation are probably brought about by changes in the balance of growth substances within the host by the pathogen. In many diseases caused by biotrophs the main obvious symptoms are abnormal growth, such as curled and distorted growth of leaves as in leaf curls of peaches and almonds, caused by *Taphrina deformans* (Plate 3e), and elongated and distorted stems, caused by downy mildews such as *Albugo candida* and *Peronospora parasitica* on *Capsella* (Shepherd's purse). Others cause galls or tumours and the stimulation of the growth of dormant buds to cause witches' brooms. Green islands often occur in tissues or organs which no longer respond to growth substance imbalance by overgrowth. These areas of delayed senescence are common around lesions caused by rusts on leaves. They suggest increased levels of cytokinins. Such green islands may also be associated with the diversion of translocates in the phloem. Cytokinins spotted onto uninfected senescing leaves often produce green islands and the host translocates towards these. These may all be considered as symptoms of disease and, too simplistically, as a consequence of the pathogen affecting the balance of growth substances in inducing translocation of host metabolites to infection sites.

Host cells around the site of a biotrophic infection show a marked increase in metabolic activity. The most marked feature is a two to four-fold increase in the rate of respiration. Such increases have also been noticed in necrotrophic infections. In host tissues infected by powdery mildews and rusts not only is there an increase in respiration rate but also a shift away from the major pathway of glycolysis and the tricarboxylic acid cycle to one involving the pentose phosphate pathway, a pathway associated with biosyntheses. Overall there is a tendency to increase the biosynthetic processes of the host to provide materials which the fungus requires, but which are also useful to the host in its attempt to resist and ward off the pathogen.

Just as the rate of respiration increases on infection, very severe depressions in the rate of photosynthesis occur in green tissues infected by biotrophs, often within a few hours of infection. As the rate of infection progresses the rate of photosynthesis continues to decline until the amount of carbon dioxide given off by the combination of host and pathogen exceeds the amount fixed. Most biotrophs not only induce the diversion of host translocates towards the infection sites but also rapidly convert them into fungal substances which the host cannot use. This ensures a concentration gradient which provides a continuous supply for the fungus and ensures that they are not re-utilized by the host. For example, host photosynthate, especially sucrose, is rapidly converted to trehalose, glycogen and mannitol.

The increased respiration of infected tissues leads to losses of carbohydrates. This loss is compounded by the effects of reduced rates of photosynthesis. These, and the diversion and conversion of host nutrients by the fungus, are the major causes of reduced growth and yield of cereals infected by powdery mildews and rusts. Healthy leaves of wheat export over 80% of their photosynthate, whereas rust-infected leaves export less than 5% and may even import photosynthate. In cereals infected by *Erysiphe graminis*, an attack on the first formed leaves, which export to the roots, restricts root development and as a consequence the number of fertile stems per plant. An attack on the upper leaves, which normally export to the inflorescences, causes poor filling and shrivelling of the grain. In some years this reduces the yield by 10–25%.

Whereas necrotrophs cause obvious signs of disease, biotrophic infections are often less obvious to the eye. They cause minimal tissue damage, but there may be morphological disturbances, they decrease yield and infected plants may senesce earlier. Often the most conspicuous early symptoms are the plants' rusted or mildewed appearance.

In biotrophic associations a well-balanced relationship is established initially between the two partners since the fungi are physiologically highly specialized in being totally dependent upon living host tissues to complete their development. This dependency is reflected in the fact that they have limited host ranges. For example, reference has already been made (p. 104) to Eriksson's discovery of a number of 'special forms' or formae speciales of *Puccinia graminis*, the black stem rust of cereals and other grasses. Where man has bred genetically stable cultivars of the host as in wheat, it is possible to subdivide each forma specialis. Not all strains within *P. graminis* f.sp. *tritici* are capable of infecting all the known cultivars of wheat to the same degree. By using a number of different cultivars of wheat as a differential host series and assessing quantitatively the degree of infection caused by the various strains, physiologic races can be recognized. In *P. graminis* f.sp. *tritici* over 200 races have been recognized. The existence of physiologic races is not a phenomenon limited to the rusts. It is a characteristic feature of specialized parasites generally, and in the fungi is well-known in both the powdery and downy mildews.

Whatever strategy a fungal pathogen uses to derive its nutrients, whether a necrotroph killing host cells in advance by pouring out cell-wall degrading enzymes and/or toxins as its hyphae penetrate between the cells of the host, or as a biotroph with similar intercellular hyphae but only tapping living cells by their haustoria, it elicits responses from the host which may or may not succeed in repulsing it. The methods by which plants resist fungal attack may be very diverse. Most plants are in fact resistant to most pathogens. Two general categories of resistance can be recognized. In the first the features which determine resistance are present before the parasite arrives on the scene. They are preformed or passive. In the second, and larger, category resistance depends on substances or structures which develop as a result of infection. They only appear when the host is provoked. They are induced or active. In the first category are found a few cases where resistance is associated with mechanical features. Thus the resistance of different species of *Berberis* to *Puccinia graminis* appears to be related to the thickness of the outer cell-wall under the cuticle. However, there is usually little correlation between cuticle and cell-wall thickness and the ease with which fungal pathogens penetrate. Suberin is not only highly resistant to fungal attack, it is also an effective mechanical barrier. Suberized cells of the endodermis may restrict the spread of vascular wilt and some root-rot fungi. The outer bark of trees, through their content of tannins, flavonoid derivatives, are chemically inhospitable and also mechanically discouraging to pathogens. Relative immunity to attack may have a chemical basis. A well-known example is found in onions in relation to 'smudge' disease, caused by *Colletotrichum circinans*. Onion varieties in which the dead, outer bulb scales are red are resistant to the disease due to the presence of the water-soluble compounds catechol and protocatechuic acid. These inhibit spore germination of the pathogen. Onions with white outer bulb scales are susceptible and lack these two compounds. Resistance is thus due to a chemical difference between the two varieties. There are many claims of phenolic substances being important in disease control. For example, quantitative differences occur in the amounts of chlorogenic acid, a phenol ester, present in the leaves of different varieties of apple. The more acid present the more resistant is the leaf of the apple to apple scab, caused by *Venturia inaequalis*. Evidence suggests that the tissues of many plants contain preformed toxic substances which are important in limiting and preventing invasion by certain pathogens.

Gaeumannomyces graminis is a soil-borne specialized necrotroph which attacks the roots of most cereals and grasses where it causes the disease known as 'take-all'. It kills virtually all the roots. It is specialized for a necrotroph in its ability to make significant growth only during the parasitic phase. It merely survives in the stubble after death of the host. Wheat and barley are susceptible to attack by most strains of the fungus, whilst oats are resistant. The reason appears to be the presence in the roots of oats of the toxic terpenoid, avenacin. Avenacin is widely fungitoxic but is particularly active against *G. graminis*. It binds to sterols in the cell membrane increasing their permeability. This absolute resistance conferred on oats by avenacin has placed a strong selection pressure on the fungus for the production of virulent mutants. The result is that a new form of *G. graminis* has evolved, *G. graminis* f.sp. *avenae* which produces the enzyme avenacinase which detoxifies the avenacin. Thus there is now a modest level of physiological specialization in *G. graminis* mediated by avenacin and avenacinase.

Resistance is largely a dynamic process. In the second category of resistance, resistance reactions are induced in the host on infection by the pathogen. Sometimes the host reacts by forming a cork layer ahead of the invader. This is so in the 'shot-hole' diseases in which the leaves look as if peppered by shot. For example, *Stigmina carpophila* causes shot-holes in species of *Prunus*. A spore infects a leaf and forms a local, circular mycelium in the tissue. However, slightly in advance of the infection, the leaf cells are stimulated to produce a layer of cork through which the fungus cannot grow. The old dead leaf areas with the fungus fall out leaving characteristic shot-holes. A barrier of cork may form below a scab lesion caused by *Venturia inaequalis* effectively sealing off the fungus and preventing further spread. Similarly haustoria penetrating cells may become enclosed by a deposit of the polysaccharide callose or lignin. It is assumed that this restricts further interchange.

Again a parasite may be restricted by the formation of gums as in the silver-leaf disease of plums. *Chondrostereum purpureum* becomes established in the wood usually through a wound. In resistant varieties, such as Pershore, gums and tannins are produced in abundance by the wood-parenchyma just ahead of the mycelium, and the advance of the fungus is physically blocked. In the susceptible Victoria plum, gum production is often inadequate and the disease spreads with disastrous results. In pine roots attacked by *Heterobasidion annosum* resin is mobilized and deposited in advance of the fungus. Resin in this case acts as both a mechanical and chemical barrier to the fungus. The insoluble resin acid impedes hyphal growth and the volatile terpene components of the resin are fungistatic. Vigorous, in contrast to suppressed, pines produce more resin and are more resistant to attack.

With many biotrophs cells of resistant plants become infected in the same way as susceptible cells but a hypersensitive reaction or response occurs. This is well-known in certain varieties of wheat resistant to *Puccinia graminis* and other rusts. In these the germ tubes from the urediniospores enter through stomatal pores and push haustoria into cells around the substomatal cavity. However, the cytoplasm of the infected cells becomes granular and discoloured and the cells rapidly die. The further growth of the biotrophic fungus is prevented because it can only derive its nutrition from living cells. The result of infection is a single dead cell or a small group of dead cells. In such biotrophs, hypersensitive death is sufficient to cause resistance. It may not be as simple as all this because similar hypersensitive death occurs on infection by some necrotrophs. The rapid browning and death of cells around the site of infection by *Botrytis fabae*, the cause of chocolate spot of broad beans, is an example (Plate 3b). Mere death of cells is unlikely to limit growth of such necrotrophs. Cell death is accompanied by a whole series of changes including shifts in oxidative metabolic pathways, accumulation of toxic substances and lignification. These partly account for some of the increase in respiration noted in and around infected tissues. It is the sum of these changes, forming part of the host's resistance and wound responses, which restricts the growth of the pathogen by not only preventing it from obtaining more nutrients in biotrophs but also poisoning it and physically boxing it in with impenetrable lignin.

A much more general type of resistance is due to the host reacting to the immediate presence of a fungus by producing a substance which, while not killing it, prevents its growth. Such a substance is known as a phytoalexin. The formation of phytoalexins is stimulated by a whole range of fungi that do not cause disease, by other micro-organisms and in response to some kind of damage, such as caused by heavy metals, to cells. The non-pathogenicity of many fungi appears to be related to the fact that they stimulate the production of phytoalexins whereas pathogenic species fail to do so or at least fail to stimulate production on such a scale as to inhibit their growth. One of the best known examples of phytoalexin is pisatin, a flavonoid, produced by the garden pea, *Pisum sativum*. This fungistatic chemical is formed freely by pea tissue, such as the inner surface of the green pods, in the presence of germinating spores of fungi which are not pathogenic to peas. On penetration sufficient pisatin is produced to inhibit and restrict these fungi to tiny discrete lesions whereas germinating spores of a fungus such as *Ascochyta pisi*, a pathogen of pea, do not, on infection, elicit phytoalexin production or immediately inactivate it.

Although much is known about certain mechanisms by which resistance to fungal attack is achieved, in most cases the exact determinants at the molecular genetic level have not been discovered. Thus the post-penetration phase is often of the nature of a struggle in which the fungus elicits responses from the host which may or may not succeed in repulsing it.

If the fungus wins this battle, the final episode in the story of a pathogenic fungus is reproduction: the production, liberation and dispersal of its spores. In many fungi only a few days elapse between infection and sporulation, as in potato blight (*Phytophthora infestans*) or in the repeating urediniospore stage of many rusts. In others, such as *Chondrostereum purpureum*, causing silver-leaf in fruit trees, many years may separate infection and basidiocarp production.

A fungal pathogen may give rise to an epidemic particularly in a crop where many genetically identical individuals are massed together. If one is susceptible all will be. In natural plant communities epidemics tend to be rare. This is because they contain a range of plant species and even in a single species there is considerable genetic variation. In most pathogens causing epidemics the life-cycle is completed in a short time and the spores are not only produced in large numbers, but are also readily dispersed. However, although a fungus needs to have these characteristics, an epidemic occurs only if the environmental conditions are conducive for the fungus. Disease expression is the result of a multiplicity of interactions and close interplay between host, parasite and the environment. Where crops are concerned man has to be added to this triumvirate. A relatively high temperature normally hastens the life-cycle and, especially when combined with high humidity, encourages spore production. Again turbulent conditions are usually favourable for spore liberation and dispersal. Finally, high humidity in combination with high temperature is of great significance in actual infection. We have already seen (p. 47) how such a fungus as *Phytophthora infestans* is well suited to produce an epidemic of blight in a potato crop when warm, damp, windy conditions prevail in late summer.

The course of many epidemics in crops have been studied. They can be described in the form of a disease progress curve (Fig. 10.5). Three phases can be recognized in the curve, an initial lag phase, an exponential phase and finally a decline phase. The initial lag is due to the small amount of inoculum present, the number of spores landing on the plants. These germinate, penetrate and there is a delay, a latent period, between infection and spore production. After a lag phase, development is at first slow but soon becomes exponential and this is checked only when the supply of uninfected host begins to decline. In potato blight, for instance, the pathogen may exhaust the supply of host by killing them all. A typical S-shaped curve is obtained, a curve similar to the growth curve of a micro-organism in a limited volume of nutrient (Fig. 10.5).

When a plant pathologist is faced with a new plant disease having a fungus consistently associated with it, his first concern is to find out if, indeed, it is the causal organism. In doing this he normally applies what are known as Koch's postulates, which were developed during the last century in relation to bacterial diseases of animals. First, the supposed pathogens must be isolated in pure culture. Secondly, it must then be inoculated into a healthy host plant and there produce the diseased condition. Finally, the pathogen must be re-isolated from the experimentally infected host and shown to be the same as the

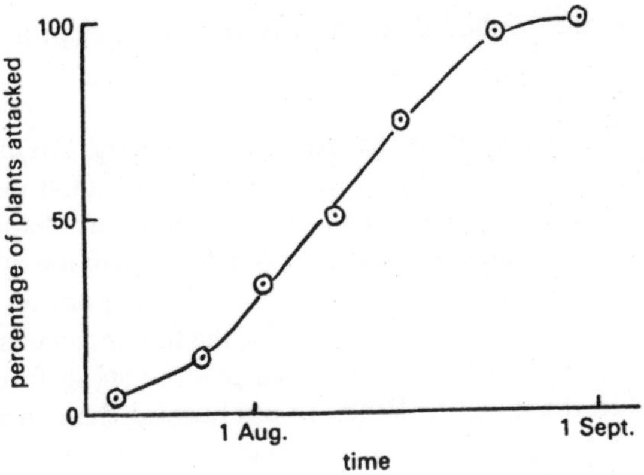

Fig. 10.5. Progress of potato blight in 117 fields in the Netherlands in 1953.

one inoculated into it. These rules can, however, be applied rigorously only when the fungus can be grown in pure culture.

Before suggesting methods of controlling a disease, the life-cycle, epidemiology and biology of the pathogen must be known in detail and the costs of the methods proposed must be carefully evaluated. In particular, the plant pathologist wants information about certain key facts: the source of spores; how and under what environmental conditions infection occurs; what is the host range, and, with a crop plant, what 'wild' species may serve as additional hosts; how the pathogen spreads from diseased to healthy plants; if and when the spores are prevalent in the air; and how and for how long the pathogen survives periods of drought and cold. Knowing all this, the plant pathologist can ascertain the weakest link in the life-cycle and apply control measures at that point. The type of host/parasite relationship is also important. In necrotrophs which kill their host rapidly, only complete prevention of infection is an adequate control measure, whereas in biotrophs where the host is debilitated sufficiently to decrease yield rather than killed, a different control strategy is possible. Any control measure which shifts the disease progress curve to the right by delaying the build-up of the epidemic can often result in much higher yields although the final level of disease is the same.

Knowing the value of the crop lost as a result of disease and the cost of the control method or methods proposed must be balanced against one another. Precision is essential in ascertaining both of these parameters. This is one reason why understanding the life-cycle of the fungus is so important but it is equally important to be able to predict when disease will occur so that control measures can be in place at the time when an epidemic is most likely. This has led to considerable efforts being put into disease forecasting.

We may now pass to a consideration of the methods used to control diseases caused by fungi. It should be emphasized that for a particular disease it may be necessary to apply a combination of control methods to ensure effective control. Even using an integrated control programme however, control is rarely complete. Further, particularly in considering the application of fungicides, the economics of the matter must constantly be kept in mind as well as possible adverse effects on the environment.

The control of diseases of plants usually concentrates on preventing infection rather than on curing plants that are already infected. An important practice is to remove, or at least reduce, the source of the initial inoculum. Thus in orchards the cutting out and burning of dead branches of apple trees killed by *Chondrostereum purpureum* and bearing basidiocarps of the fungus, or of cankered branches producing spores of *Nectria galligena*, has this end in view. In apple scab, caused by *Venturia inaequalis*, the first infections in May are mainly from ascospores produced in ascocarps developing from dead, over-wintered leaves on the ground. Gathering and burning these in autumn, or ploughing them into the soil, greatly reduces the source of primary infection in the late spring. Another method of dealing with an over-wintering stage is by crop rotation. This is valuable in the control of ergot. In rye infected by *Claviceps* (Fig. 10.1) the sclerotia fall to the ground and remain dormant over the winter, but may give rise to stalked perithecial stromata in the following summer from which ascospores are discharged. Rotation helps by denying the fungus an immediate supply of host plants, and in this connection it must be remembered that the great majority of spores travel only short distances (Fig. 7.3A). Rotation is, however, effective only if the resting stage persists for a single winter. In some fungi it remains dormant but viable for years. This is so in *Synchytrium endobioticum* which causes wart-disease of potatoes (Fig. 10.1). Rotation is important in dealing with take-all of wheat due to *Gaeumannomyces graminis* which persists over winter on stubble ploughed into the soil. The disease potential of the fungus can also be reduced by encouraging the competitive activities of normal soil saprotrophs by ploughing in a green manure such as mustard planted immediately after harvest. The plant pathologist concerned with soil-borne disease is dealing with ecological problems in which host, parasite and other soil organisms are interacting in a complex environment.

In a heteroecious rust the source of infection can be removed by eradicating the alternate host. This has already been discussed in connection with the black stem rust of wheat (p. 105), but rooting out barberry succeeds in preventing infection of wheat only where the repeating urediniospores of *Puccinia graminis* cannot survive the winter. In rusts without a repeating spore stage, removal of one host, where this is possible, gives complete control of rust on the other.

Using chemicals to control plant diseases is one of the oldest and still most widely practised methods of disease control. Essentially three methods are possible, using chemicals as eradicants, as protective or as systemic fungicides. Many fungi produce spores which persist in the soil and then attack plants. One method of combating soil-borne fungi is soil sterilization. This is often not practicable for field crops because of the expense although it has been used in the field to prevent infection of high value crops. The most widely used sterilant in greenhouse and small field plots is the volatile and very poisonous methyl bromide with the object of killing pathogenic species of *Fusarium*, *Pythium* and *Verticillium*. It leaves no residues but apart from its high application cost it kills everything in the soil upsetting the delicately balanced ecosystem there.

Prevention of infection by the application of protective fungicides plays a central role in the control of diseases. Two inorganic types have been extensively used, namely sulphur or lime-sulphur (a mixture of polysulphides and thiosulphates of calcium) and copper compounds. Compounds of mercury were also formerly used but, because of their extreme toxicity to animals, they are generally not used today for environmental reasons. Fungicides based on sulphur have proved especially effective against powdery mildews, such as *Uncinula necator* which attacks grapes, rusts and apple scab. Copper-based fungicides are noteworthy in connection with their early use against potato blight. Bordeaux mixture was the first chemical fungicide and is still used today. Solutions of copper ions are toxic to plants and insoluble copper salts are more widely used. Protecting a crop with a fungicide involves using a spraying apparatus which deposits the fungicide evenly over the whole shoot system. The idea is that the fungitoxic chemical forms a deposit on the leaves and kills the spores as they germinate but before they penetrate. Further, this must be done at the right time, namely just before the spores of the pathogen become prevalent.

In more recent years such fungicides have tended to be replaced by heavy metal organic ones. Dithiocarbamates are now widely used. Maneb and Zineb, manganese and zinc salts of dithiocarbamate, are effective against potato blight and rusts.

Protective fungicides are also used as dressings to control seed-borne diseases. In the control of covered-smut of barley and oats, caused by *Ustilago hordei*, and other seed-borne diseases, such as those caused by *Helminthosporium*, the seed is usually dressed with an organo-sulphur fungicide in a powder form. These fungicides only control the pathogen if it is present on the surface of the grain killing the spores as they germinate. Some fungi are more deep-seated with a dormant mycelium within the grain. This is true of loose smut of wheat and barley, caused by *Ustilago nuda*. This is not eliminated by the normal practice of seed dressing. It was formerly controlled by a rather difficult hot-water treatment which depended on the fungus being more sensitive to heat than the cereal embryo. Recently, however, this smut has been controlled by systemic fungicides.

All these fungicides are very useful and are widely used but they have a number of disadvantages. They may be toxic to plants and animals. They are eventually washed off by rain, although spreading and sticking additives are incorporated in their formulation. As the plant grows new unprotected surfaces are exposed to the pathogen. Several applications are necessary throughout the season. The cost of application is high and there may be machinery damage to crops. The major problem, however, is that the fungicides have no eradicant effect. Once the disease is established in the crop, the fungicide has no effect. This has led to the development of a number of systemic fungicides over the last 30 years. These include derivatives of benzimidazole, pyrimidine and morpholine. These substances, toxic to fungi but not to their host, are not only absorbed into the threatened plant, but may also be translocated within them to all parts. They may be applied in solution to the roots, as a spray to the shoots or as a seed dressing. One of the most successful has been benomyl. It has been used as a seed dressing to control *U. nuda*, as a dip for bulbs to control *Fusarium* and *Penicillium* rots, as a foliar spray to control powdery mildews and apple scab and as a root drench to control vascular wilts. It has also been injected into elms to render them resistant to Dutch elm disease.

Systemic fungicides have a number of advantages over protective ones. They kill the fungus after infection. Because they are taken up, they tend to persist longer, so fewer applications are necessary. As a group they are less toxic to animals and plants. There are, however, two very important disadvantages. Their cost of development is very high, so they are expensive. Most of the copper, sulphur and dithiocarbamate fungicides have remained effective over the years. This is probably because they are more general poisons affecting a number of central metabolic reactions in the fungus. Stable resistance to systemic fungicides, such as benomyl, has arisen in such pathogens as *Botrytis cinerea* and *Venturia inaequalis*. This is because their toxic action is site-specific in the pathogen, rather than multi-site as in protective fungicides. Resistant strains have arisen either by the selection of resistant individuals or as a result of single major gene mutations. Such fungicides should be used in various combinations or alternated to minimize the possibility of resistance developing to any one.

In the control of diseases of crop plants, the breeding of resistant varieties plays a major part. The genetic make-up of both the host and the parasite must be considered. In the middle of the present century a genetical study of flax in relation to infection by the rust *Melampsora lini* led to the gene-for-gene concept: 'for each gene conditioning rust reaction in the host there is a specific gene conditioning pathogenicity in the parasite'. Any genetic system determining virulence in the pathogen will be paralleled by genes conferring resistance in the host. In the flax–rust combination it was found that there are two genes in the host which, when represented by the dominant alleles, confer resistance on the host matched by genes in the parasite which, when represented by recessive alleles, develop virulence. A resistant reaction only occurs where a dominant allele for resistance in the host plant interacts with a dominant allele for avirulence in the pathogen (Fig. 10.6). The genetics of resistance has now been explored in a wide range of rust and powdery mildews and in potato blight and tomato leaf-mould caused by *Cladosporium fulvum*. Often the number of resistance genes in the host matched by virulence genes in the pathogen is quite large, as many as 11 in the case of potato and *Phytophthora infestans*. This type of resistance under single, dominant gene control is called major gene or race specific resistance. When the fungus infects a resistant host, it is rapidly killed by a hypersensitivity response and it never goes on to reproduce. Such resistance is easily handled by breeders since its inheritance obeys Mendel's laws and it is very effective, as the fungus is killed. Screening of seedlings for resistance is simple and rapid. However, the introduction of a new resistance gene in the host into a crop as a new resistant

		Host genotype	
		R	r
Pathogen genotype	Av	Resistant	Susceptible
	av	Susceptible	Susceptible

Fig. 10.6. Interactions between alleles of a host resistance gene and the corresponding pathogen gene for virulence. Resistance (R) and avirulence (Av) are dominant.

cultivar is almost invariably followed by a new race of the pathogen capable of overcoming that resistance. A single gene mutation in the fungus leads to a breakdown of that resistance. Since resistance is all or nothing, there is little or no competition to limit the spread of such mutants on a susceptible host. A series of physiologic races of the pathogen soon build up, as seen in black-stem rust, each race capable of overcoming different resistance genes in the host. Very few resistance genes remain effective for more than a few years.

However, all is not completely lost and breeding for major gene resistance is still an important component of disease control. Plant breeders have adopted three approaches to combat the problem. Using a set of differential host cultivars, it is possible to conduct race surveys of any particular pathogen to find out what races are present and their frequency in the entire domain from which infection could occur. Advice is then given on suitable cultivars to plant and those not to plant. To foretell and deploy resistance accordingly is to keep almost one step ahead of the pathogen. Such race surveys are carried out across Europe annually for yellow and brown rust and powdery mildew of cereals.

Another approach, especially in cereal growing, using the same information is to recommend sowing mixed cultivars. These are mixtures, mechanical blends, of three or four cultivars, differing in major resistance genes but similar agronomically. Even if all the virulence genes are present in the population of the fungus to overcome the resistance genes present, the epidemic builds up more slowly than it would have done in a crop of a single cultivar. There are a number of possible explanations for this. The area of susceptible tissue available in the crop to each virulence gene is reduced. This reduces the number of infective spores in the next generation. The distance the inoculum has to travel between plants is also increased and the movement of inoculum from one susceptible plant to another is hindered by intervening resistant plants. Spores are to some extent 'filtered out'. The net effect is to reduce the rate at which the epidemic builds up. Although this is a cheap and simple means to curtail disease and allows the tailoring of mixtures to suit local needs, it has not been widely taken up because of market forces.

The third approach is to attempt to produce multiline cultivars. In a multiline cultivar all the plants have an identical genetic background except that they may differ in the resistance genes they possess at one or a few loci. The plants are thus identical in agronomic characters but differ only in race-specific resistance. These again slow down the development of an epidemic and reduce the selection pressure on the pathogen to overcome resistance. However, these isogenic lines are expensive and slow to produce because of the repeated backcrossing that has to occur to retain the agronomic qualities. Their other main problem is that, by only differing in resistance genes for one pathogen, they provide a uniform genetic background with respect to non-target pathogens.

The basic idea behind all these strategies is diversification in resistance to avoid exposure of a single resistance gene over large areas. The 'boom and bust' cycle, the increased use of a new cultivar with major gene resistance followed by its withdrawal after virulence has developed in the pathogen, has led to an increased interest in race non-specific, polygenic or multigene resistance. In this type of resistance all cultivars of a plant react similarly to all races of a particular pathogen and show a general level of resistance, which changes little with time. It is controlled by a large number of genes. It is not affected by changes in the virulence genes of the pathogen. The net effect of the host is not to kill the pathogen as soon as it infects but rather to reduce its growth. This results in less drain on the host, the fungus takes longer to reproduce and fewer spores are produced. The result is a slowing down of the rate of build up of the epidemic. There is no selection pressure in favour of virulent mutants of the pathogen because it is already present occupying infection sites. Mutants do not therefore build up and there are no physiologic races. The major disadvantages are that because the pathogen is not killed on infection, some losses do occur and it is difficult to handle because of the complex genetics. It is not as widely used as it should be.

After their arrival on the surface of the host, but prior to infection, pathogens may be susceptible to competition or antagonism by other micro-organisms. These latter are exerting a degree of biological control. Even dead spores of some species of *Trichoderma* will prevent colonization of a pruning wound on a plum tree by *Chondrostereum purpureum*. Once the latter is established in the wound, the *Trichoderma* spores are ineffective.

The principle of biological control sounds easy but very few methods so far devised have been commercially successful. The best known example is that of the control of *Heterobasidion annosum* on pines. In its parasitic attack, *H. annosum* depends on first becoming established on the exposed surfaces of freshly cut stumps resulting from thinning in the young forest. From these invaded stumps, the fungus spreads to surrounding trees by root contact. Good control is achieved by spraying a suspension of conidia of *Peniophora gigantea*, a saprotrophic wood-rotting Basidiomycotina onto the cut surfaces of the stumps immediately after felling. This harmless saprotroph rapidly becomes established and easily wins in competition for nutrients in the wood with the mycelium of *H. annosum* which may develop on the stumps from airborne basidiospores. It also antagonizes *H. annosum* by a form of contact antibiosis known as hyphal interference.

Finally, in efforts to control plant disease, it is occasionally possible to take steps to prevent their spread. This is, however, not practicable if the pathogen is a fungus producing large numbers of airborne spores. When, however, there is no airborne phase, as, for example, in *Synchytrium endobioticum* which causes the serious wart-disease of potatoes, spread can be prevented. This disease is contained by allowing only immune varieties of potato to be grown in infected regions of the country and by forbidding movements of tubers out of these areas.

Most countries attempt to prevent the accidental introduction of new diseases by appropriate quarantine regulations. For example the USA requires a special permit for the import of any kind of plant material except seeds.

The parasitic activity of fungi extends to organisms other than plants. Some attack insects. *Entomophthora muscae*, a member of the Zygomycotina, has already been discussed as a parasite of flies (Chapter 3). In the Ascomycotina, *Cordyceps* is an equally well-known parasitic genus. Some 150 species are known. Most attack Lepidoptera and Coleoptera, but in the tropics some attack adult spiders. *Cordyceps militaris* produces orange-coloured, stalked, club-shaped stromata, 10–40 mm tall, bearing numerous perithecia immersed in the swollen apex. These arise in the autumn from mummified larvae, pupae or adults in the soil. Ascospores are released and, on infection, short, cylindrical hyphal bodies develop in the haemocoel of the insect. These increase by budding and become distributed throughout the body. Death occurs after about 5–7 days. After death, hyphae develop from the hyphal bodies and the dead insect becomes plugged with a dense mass of fungal tissue. At this stage it is virtually mummified. Its tissues are extremely resistant to decay due to the production of the antimicrobial nucleotide cordycepin by the fungus. This ensures that other saprotrophic micro-organisms in the soil do not plunder the reserves required to produce the stalked ascocarps.

Fungi may also attack other very small animals. Reference has already been made to those conidial fungi that trap and consume eelworms (p. 116). There are also some interesting Zygomycotina which parasitize amoebae in the soil. At the other end of the scale fish, birds and mammals are sometimes subject to fungal attack. In a later chapter, human mycoses will be briefly considered.

Fungi not infrequently attack other fungi. Some chytrids parasitize water moulds and even other chytrids. Certain of the Mucorales regularly infect members of the same order. Some mucoraceous fungi find their hosts among the higher fungi. For example *Spinellus* is frequently to be seen on basidiocarps of the agaric genera *Mycena* and *Collybia*. Also relatively large

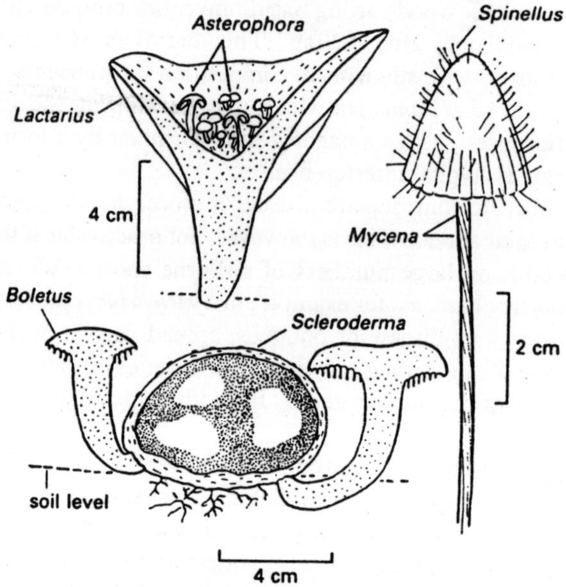

Fig. 10.7. Fungi parasitizing fungi. *Asterophora parasitica* growing on *Lactarius vellereus; Boletus parasiticus* on *Scleroderma citrinum*; and *Spinellus* sp. on *Mycena polygramma*.

fungi may attack other higher fungi. *Boletus parasiticus* grows on the earth-ball (*Scleroderma*), and the little toadstool *Asterophora parasitica* is a common parasite of the much larger agarics *Lactarius vellereus* and *Russula nigricans* (Fig. 10.7). These are only a few examples of parasitic attack by fungi on other members of the same kingdom.

Fungi as mutualistic symbionts 11

A wide range of fungi live in mutualistic associations with plants. The best known of these symbioses are mycorrhizas and lichens. Several quite distinct symbioses between mycelial fungi and the roots of seed plants are grouped together as mycorrhizal associations. In association the two partners form a unique absorbing structure, a 'fungus root'. These associations have several features in common: they are relatively constant; infection is the norm; most plants are mycorrhizal most of the time; the fungi are non-pathogenic; growth of the seed plant is frequently, if not invariably, enhanced by the development of mycorrhiza; both partners benefit from the association; and they are thus mutualistic symbiotic associations.

Traditionally mycorrhizas have been divided on structural differences into ectomycorrhizas and endomycorrhizas. This distinction depends upon whether the bulk of the fungus is outside or inside the root. Only some 3% of seed plants have ectomycorrhizas or sheathing mycorrhizas. They are most common in North Temperate forest trees, especially beech, birch, oak and pine but are also well-represented in some tropical families. It has been noted (Chapter 9) that each type of woodland, beech, birch, oak or pine tends to have a distinctive assemblage of larger fungi growing on the ground. In small part this is due to the preference of certain species for a particular kind of leaf litter. In the main, however, the agarics and boletes growing on the forest floor are not litter decomposers. Fleshy toadstools, particularly those in such genera as *Lactarius*, *Russula*, *Amanita* and *Boletus*, only occur under certain kinds of tree. Thus *Lactarius quietus* is found under oak (*Quercus*), *L. turpis* under birch (*Betula*), *Russula fellea* under beech (*Fagus*), *Suillus grevillei* under larch (*Larix*) and a few other conifers and *Russula emetica* under pine (*Pinus*). *Amanita muscaria*, the fly agaric, occurs only under birch and pine. These restrictions relate to the fact that these fungi form mycorrhizas with these particular trees. The relationship between fungus and tree has been particularly studied in beech and pine.

In a beechwood, if the deep leaf-litter is brushed aside, the ultimate fine lateral rootlets can be seen. These are differentiated into 'long' and 'short' roots. The long roots are the extending roots and grow indefinitely. The absorbing roots are their branches, the short roots. These are modified into

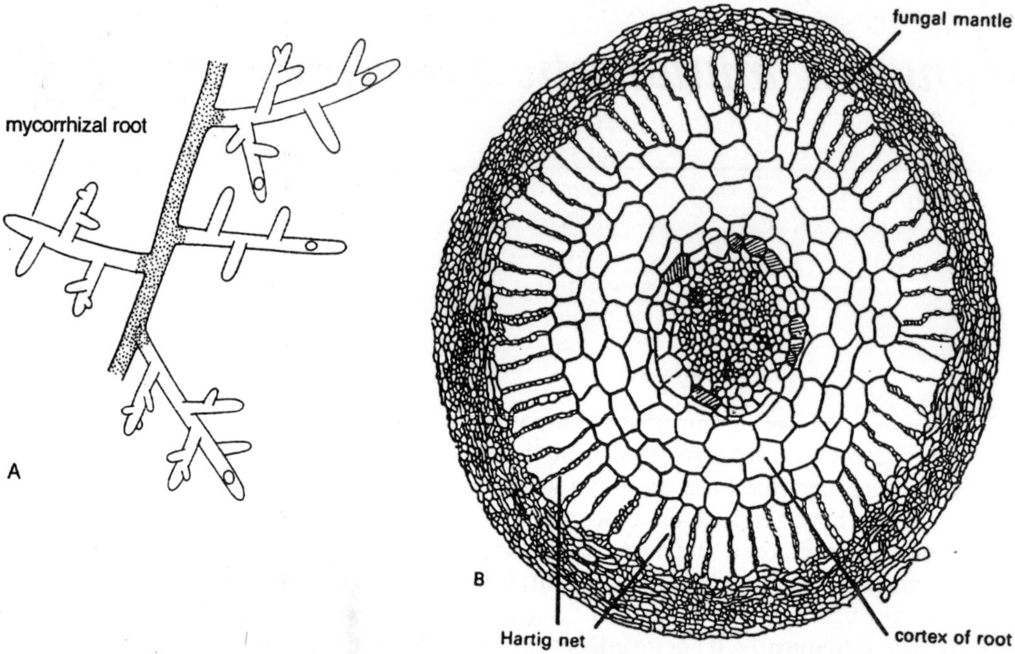

Fig. 11.1 Beech (*Fagus*) mycorrhiza. A, portion of a root from the leaf-litter with highly branched mycorrhizae (shown white). B, transverse section of a mycorrhizal root showing the fungal mantle and the intercellular hyphae between the outermost root cells (× 107). (After Clowes)

mycorrhizas. They are of limited growth, shorter lived, thicker, more branched, almost coralloid in appearance and a rich golden-brown in colour (Plate 4a). A transverse section shows that the host root is surrounded by a dense sheath of fungal hyphae, comprising some 20–30% of the total volume of the root, with hyphae penetrating between the outermost cell layer of the root (Fig. 11.1 and Plate 4b). From the outside of the sheath hyphae, or mycelial strands in some, spread varying distances out into the surrounding soil and litter.

Most ectomycorrhizal fungi are Basidiomycotina, although a few Ascomycotina are also involved in such associations. Endomycorrhizas are more common, especially the vesicular-arbuscular mycorrhizas produced by non-septate fungi from the Zygomycotina Endogonales. The mycorrhizas are so-called because of the two characteristic structures found in roots with this type of infection – vesicles and arbuscules (Fig. 11.2). They are found in virtually all groups of plants, including mosses and ferns, and are general in such seed plant families as those including the grasses, the palms, roses and peas. All of these include many important crop plants. Most plants possess this type of mycorrhiza. In this type all that is seen externally is a loose and very sparse network of irregular, rather thick-walled and non-septate hyphae. From this network hyphae run several centimetres out into the soil. At an infection point

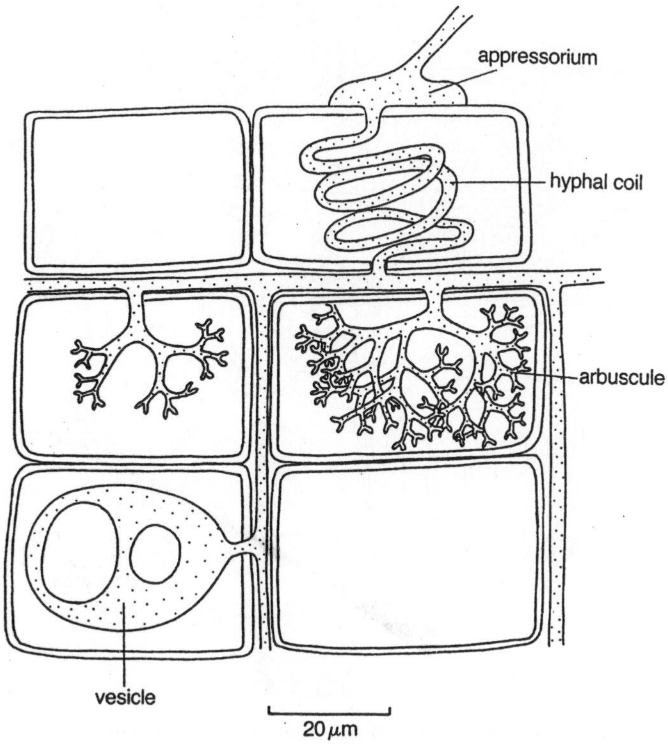

Fig. 11.2. Vesicular-arbuscular mycorrhiza. Diagram of portion of cortex of a root showing appressorium, hyphal coil in epidermal cell, intercellular hyphae, arbuscules and a vesicle.

on a root an irregular disc- or lozenge-shaped appressorium is formed. An intracellular hyphal coil is often formed in the epidermal cell beneath and from this hyphae grow out and penetrate between the cortical cells of the root. The meristem of the root is not infected but the hyphae grow out and permeate the entire cortex. Inside the host cells arbuscules are produced, so-called because they resemble little trees. The cell-wall is penetrated by a trunk hypha which inside the wall branches dichotomously invaginating the cell membrane. The ultimate branches may be so fine that they are difficult to resolve. They often fill the host cell. They are complex haustoria but, unlike haustoria of biotrophic pathogens, they are lysed after about 4 days. Lipid droplets are released into the host cells. Intercellular hyphae are also lysed. Lipid droplets may appear in these and they may become septate. Vesicles are formed at the tips of hyphae, which may be intercellular or intracellular. They are ovoid or spherical and contain one or more lipid droplets and probably act as storage structures. In the soil, hyphae produce spores. The most common type are very large, up to 250 μm in diameter, golden-brown to black, thick-walled, balloon-like chlamydospores (Plate 4c). These have large food reserves and persist in the soil for long periods. They are probably stimulated to germinate

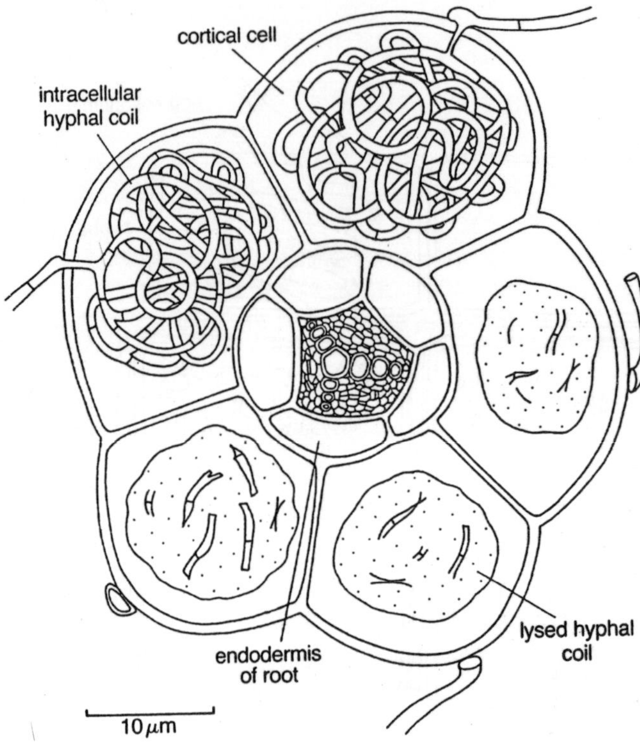

Fig. 11.3. Ling (*Calluna*) mycorrhiza. Transverse section of a hair-root showing two cells with intracellular hyphal coils and three cells where the coils have undergone lysis to become structureless amorphous masses.

and infect roots by root exudates. Infection is also common by hyphae growing out from an infected root contacting an uninfected one.

There are a variety of types of endomycorrhizas. Heather (*Erica*), ling (*Calluna*), rhododendrons (*Rhododendron*) and bilberry (*Vaccinium*) which grow on nutrient poor, acid, peaty humus soils are virtually always mycorrhizal. They have a very dense root system ending in very fine, branched, absorbing 'hair roots'. These rootlets have a very narrow central stele surrounded by a cortex of usually one layer of parenchymatous cells (Fig. 11.3). These rootlets are covered by a very sparse weft of dark brown, septate hyphae. From this weft branches penetrate the cortical cells forming very compact hyphal coils invaginating and enclosed by the host's cell membrane. After a period of time the intracellular coils are also lysed. There is a loss of filamentous structure. The coils become an amorphous structureless mass and may disappear altogether. This lysis is comparable with that in the arbuscules of vesicular-arbuscular mycorrhizas and, as we shall see, in the mycorrhizas of orchids. The fungi involved in these so-called ericoid mycorrhizas are Discomycetes in the main.

Although on good, nutrient rich soils plants can grow without a fungal partner, it seems that on nutrient poor soils survival depends on the mycorrhizal relationship. Mycorrhizal infection tends to be more intense under conditions of mineral nutrient deficiency or imbalance in the soil. It is in such conditions as relative mineral deficiency or where there is competition from other plants, indeed in most natural habitats, that mycorrhizal infection increases the growth of plants. In all these systems mycorrhizal compared with non-mycorrhizal plants show greater uptake of mineral ions such as those of nitrogen, potassium and especially phosphorus and also of other more minor mineral nutrients. This is achieved by a number of means but can be largely explained on the basis of the increased absorbing area of the absorbing system in mycorrhizal roots. For example, in sheathing mycorrhizal systems the absorbing area is increased by the greater diameter of absorbing roots, their increase in number by the stimulation of branching, the extension of hypha, often as mycelial strands, from the sheath well out into the soil or leaf-litter, although this is partially offset by the fact that mycorrhizal roots lack root hairs, and that mycorrhizal roots live longer and therefore absorb longer. Such is also true of vesicular-arbuscular mycorrhizal systems. Phosphate ions in the soil are very firmly adsorbed onto the clay micelles and have a very slow diffusion coefficient. There is likely to be a marked depletion of phosphate in solution immediately around the roots. In non-mycorrhizal plants this is about 1–2 mm. In vesicular-arbuscular mycorrhizal roots this zone extends up to 100 mm from the roots depending how far the hyphae extend out permeating the soil. Mycorrhizal roots thus tap a larger volume of soil for the exchangeable mineral ions. There is also some evidence that the hyphae of the mycorrhizal fungi are more efficient absorbers than the roots themselves and are able to extract ions from soils below the minimum threshold available to the roots. Thus the enhanced mineral ion uptake by mycorrhizal roots may depend both on a more complete exploration of the soil by the hyphae and a more complete exhaustion of the soil.

There is also good evidence that they may make available organic nitrogen and perhaps phosphorus which are not available to non-mycorrhizal plants. In the acid, peaty soils in which plants such as *Calluna* and *Erica* grow nitrification rates are very low and the available nitrogen source is ammonium ions which are not only in low concentrations but also relatively immobile. Most of the nitrogen, well over 70%, is present in the organic form and unavailable to plant roots. Mycorrhizal plants grown in such soils are larger and healthier and their nitrogen, and phosphorus, content is significantly higher. It has been shown that mycorrhizal plants can use amino-acids derived from humified materials and from leaf leachates in particular, whereas in non-mycorrhizal plants this ability is very restricted. The mycorrhizal fungi can utilize such simple organic compounds and make them available to the plant. More recently it has been shown that most mycorrhizal fungi produce extracellular proteases so can cleave peptides and proteins to amino-acids and take these up.

In short-term experiments with sheathing mycorrhizal systems, the primary site of absorption is the fungal sheath. About 90% of the phosphate taken up in the first few hours is retained in the sheath. It acts as a primary reservoir of mineral ions, then releases these to the host in periods of deficiency.

Members of the Ericaceae, such as *Calluna* and *Vaccinium*, are also very successful colonizers of natural and man-made environments which contain very high levels of toxic heavy metals. *Calluna vulgaris* is frequently the dominant and often the only colonizer of coal tip spoils and *Vaccinium macrocarpon* is prominent on soils contaminated by metal smelting. Enhanced uptake under these circumstances would be physiologically damaging rather than beneficial. It has been found that the concentration of heavy metals is lower in the shoots of mycorrhizal than non-mycorrhizal plants and the concentration is lower in the shoots than in the roots of mycorrhizal plants. It appears that heavy metals are complexed in the hyphal walls in mycorrhizal roots which facilitates exclusion of metals from the shoots and avoids metal toxicity. The fungi themselves appear to be tolerant of heavy metals. In sheathing mycorrhizas heavy metals appear to be sequestered in the sheath.

Mycorrhizal plants benefit in a number of other ways from their symbiotic fungi, especially in their water relations. We have seen that mycorrhizal plants because of their extensive hyphal system tap a larger volume of soil for its exchangeable mineral ions. Hyphae can also absorb and conduct water over long distances. As with mineral ions they can tap the same but much larger pool of water in the soil and as in many soils water, like minerals ions, is in short supply, the host plant is bound to benefit under stress where water supply to the non-mycorrhizal plants is restricted.

Most mycorrhizal fungi appear to be obligate biotrophs in nature. They are only found growing with their hosts. They benefit by obtaining their organic nutrients from their host plant. They are entirely dependent upon their host for these. They require simple sugars and cannot use polymers such as cellulose. In sheathing mycorrhizas, for instance, the sheath forms a sink for host photosynthate. Sucrose is rapidly translocated to the roots and converted to the fungal carbohydrates trehalose and glycogen and the polyhydric alcohol mannitol. None of these can be effectively reabsorbed and utilized by the host. This diversion and conversion parallels that seen in biotrophic pathogens, such as the rusts. The large and organized sheath is a conspicuous feature of ectomycorrhizas, which are most common in our North Temperate trees. In beech, it comprises some 35–40% of the dry mass of the mycorrhizas and is responsible for about half of their carbon dioxide emission. It must be a very expensive structure to upkeep and so must have some selective value. It is a storage structure for fungal carbohydrates and mineral ions. A carbohydrate storage facility may have a selective advantage for fungi which produce a long-lived mycelium and seasonally produce large reproductive structures, such as basidiocarps. They can store up carbohydrates in the sheath when the host is most active in photosynthesis in the spring and summer and utilize this store in autumn and after leaf-fall. This is the season when conditions are most opportune for the development of agaric-type basidiocarps and when their spores can most effectively be produced and dispersed. Most agarics are fleshy, relying on turgor for support. They are unable to withstand the dry days of summer and succumb to the frosts of late autumn and winter. The availability of minerals in the litter and soil is also seasonal. After leaf-fall in the autumn, the litter is moist and microbial and micro-faunal activity is high. Minerals will be made available for absorption by leaching from the litter and mineralization. In the dry summer however both of these processes will be depressed. Efficient absorption by the fungi in times of plenty or availability and storage for release to the host in its growing season when they are scarce in the environment has obvious merits. Ectomycorrhizas thus appear to be adaptations mainly shown by trees to seasonal climate, particularly those of the North Temperate zone, where, to some extent, the host and fungus grow and reproduce out of phase with one another. In the more equable Wet Tropics, the trees and their mycorrhizal fungi grow and reproduce throughout the year. Most, but not all, the trees have vesicular arbuscular-mycorrhizal fungi.

Unlike most biotrophic pathogenic fungi, many mycorrhizal fungi show a remarkable lack of host specificity. In many plant communities mycorrhizal infection arises when uninfected roots are contacted by hyphae spreading into the soil from infected roots. Since any fungus may be capable of infecting a number of different plant species, a series of plants in different species may be connected by the hyphae of their common fungus. It has been shown that this can lead to inter-plant transfer of both mineral ions and carbon compounds. This may have far-reaching ecological consequences.

It is well-known that seedlings of some woodland and grassland habitats can survive prolonged periods of exposure to shade. Such shade tolerant species would be expected to be mycorrhizal as seedlings. Early mycorrhizal infection would be vital in providing a necessary enhanced mineral nutrient supply for rapid establishment but the carbon drain to provide such enhancement could be critically disadvantageous to growth. Direct transfer of photosynthate via a common mycorrhizal fungus from more fully illuminated neighbours or over-storey plants may sustain seedling infection. Thus during the critical establishment phase the whole seedling may live for a period, at least, as a partial heterotroph depending upon its more fortunate neighbours to which it is connected by a fungal bridge. In many plant communities, seed germination takes place in very small gaps in the established vegetation. In chalk grasslands, for instance, in members of the Gentianaceae, such as *Blackstonia perfoliata*, seedlings arise in small gaps, 10–50 mm in diameter, between established plants. Germination in such a restricted root space might be expected to lead to the early development of mineral nutrient deficiencies, but the seedlings very rapidly become infected by hyphae spreading out from surrounding unrelated infected plants and the seedlings soon benefit from the transfer of mineral nutrients. The seeds of these species germinate in the autumn. This is the time when many of the adult plants, such as the grasses, in the community have peak levels of nitrogen, potassium and phosphorus after their growth in the spring and summer. These are well-equipped to act as donors.

In nature orchids are dependent on mycorrhizal fungi at some stage of their development. In those with green leaves, dependence may be limited to germination and the seedling stage, whereas in those lacking chlorophyll (e.g. *Corallorhiza*, *Galeola* and *Neottia*) reliance on mycorrhizal fungi is obligatory for all their lives. The fungi concerned are mainly species of *Rhizoctonia*, a genus of Basidiomycotina, found normally only in the mycelial state. The mycelium consists of branched, septate, brown hyphae and small black sclerotia are produced. Nutritionally these fungi are much less exacting than are those involved in other mycorrhizal systems, being able to utilize both cellulose and lignin. The majority of species of *Rhizoctonia* are known to be capable of existing as free-living competitive saprotrophs in the soil. Many have been demonstrated to be aggressive necrotrophic parasites of other hosts. They cause root diseases of cereals, tomatoes, lettuces and other crop plants and 'damping-off' diseases of seedlings such as those of cress and swedes.

Orchids produce millions of minute seeds, each of which contains an undifferentiated embryo of only a few cells and weighs less than 0.01 mg. There is no food reserve. The majority are incapable of successful germination unless they are provided with an external source of soluble carbohydrate, such as glucose. In moist soil the seed absorbs water, swells slightly, ruptures the seed coat and produces a few epidermal hairs. It develops no further unless it becomes infected by an appropriate mycorrhizal fungus. Infection is followed by an immediate marked stimulation of the growth of the seedling. It differentiates a central conducting strand and eventually an apical group of leaves and roots from the axis near the leaf bases. The fungus becomes restricted to the outer parenchymatous cells surrounding the inner vascular strand with hyphae passing out into the soil. As the young roots develop, they in turn become infected from the fungus in the soil. The fungus only enters the parenchymatous cortical cells of the root and it forms dense intracellular coils or tangles of septate hyphae (Fig. 11.4). The cell membrane of the host cells is not penetrated but becomes extensively invaginated as the coils progressively permeate the cell thus creating a large area of host–fungus interface. With time these hyphae swell, lose their contents and are apparently lysed by the host into a structureless, disorganized mass. The whole situation suggested a precarious balanced parasitism, and, indeed, there is evidence that many orchid seedlings perish by failing to contain the fungus. It grows throughout the seedling killing it as a necrotrophic parasite. In vigorously growing seedlings a balanced relationship is established and there is good evidence that the fungus utilizes complex carbon sources in the soil and makes the hydrolytic products available to the young orchid. It seems likely that the hyphae in the soil which pass into the host may translocate organic carbon in the form of the fungal sugar trehalose to the living fungal coils where there is direct transfer to the host. Trehalose forms an adequate carbon source for the asymbiotic growth of many orchid seedlings. In addition the orchid may obtain some organic nutrients by the periodic lysis of the hyphal coils. Lysis is, however, more likely to be an active general defence mechanism against complete parasitic invasion. The extent and rate of movement of organic carbon from fungus to host would suggest that translocation is the major pathway for transfer.

Any orchid relies upon its fungus for its organic nutrients until it is photosynthetic. It may be some considerable time before the first photosynthetic leaf is produced – 2, 10 or more years. So all have a prolonged and

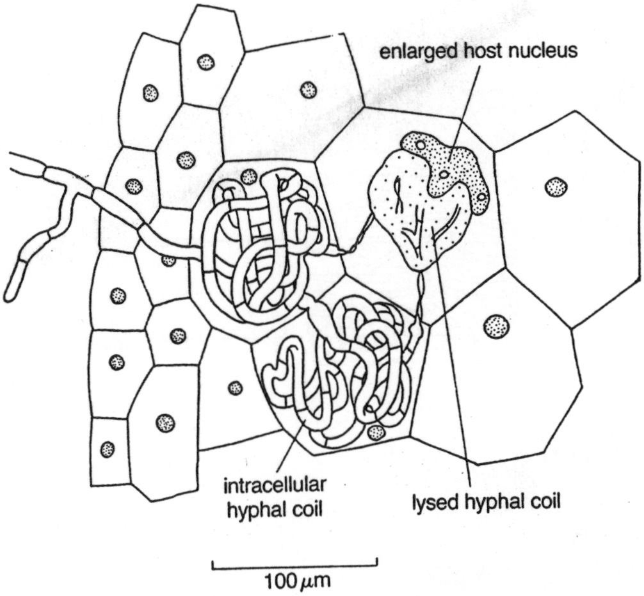

Fig. 11.4. Orchid mycorrhiza. Diagram of part of a transverse section of a root showing intact and lysed intracellular coils.

some a very prolonged phase in the seedling stage when they are non-green, and therefore heterotrophic, subterranean and obligatorily mycorrhizal. During this phase in the balanced association there is a net movement of organic carbon, and mineral nutrients, from the fungus to the orchid. The fungus receives no benefits. At this stage this association must be regarded as a parasitic symbiosis with the orchid parasitizing the fungus. The nature of the relationship between the adult green orchid and its mycorrhizal fungus is less clear. In many orchids the cortex of their roots becomes infected as they develop each year as in other mycorrhizal systems, although no movement of photosynthate from the host to the fungus has ever been demonstrated. Nor is there any evidence that the orchid receives mineral nutrients from the fungus. There is much still to be learnt about the association. Perhaps root infection is inevitable because of the aggressiveness of the fungi as root parasites but the possibility of there being a reciprocal exchange of carbon from the orchid to the fungus cannot be definitely ruled out.

Lichens are the best example of two organisms living together. A lichen is a dual organism in which a fungus is intimately associated with an alga or a cyanobacterium. Their physiological interactions result in a new distinct morphological entity which is a self-reproducing functional unit. It behaves as a single organism and has no obvious similarities to either of its components. If only a few lichens existed, they would simply be incorporated in fungal classification as interesting species living on algae or cyanobacteria. However, lichens are so numerous, with over 13 500 species, and they tend to occupy such distinctive ecological niches, that they are placed in a separate taxon.

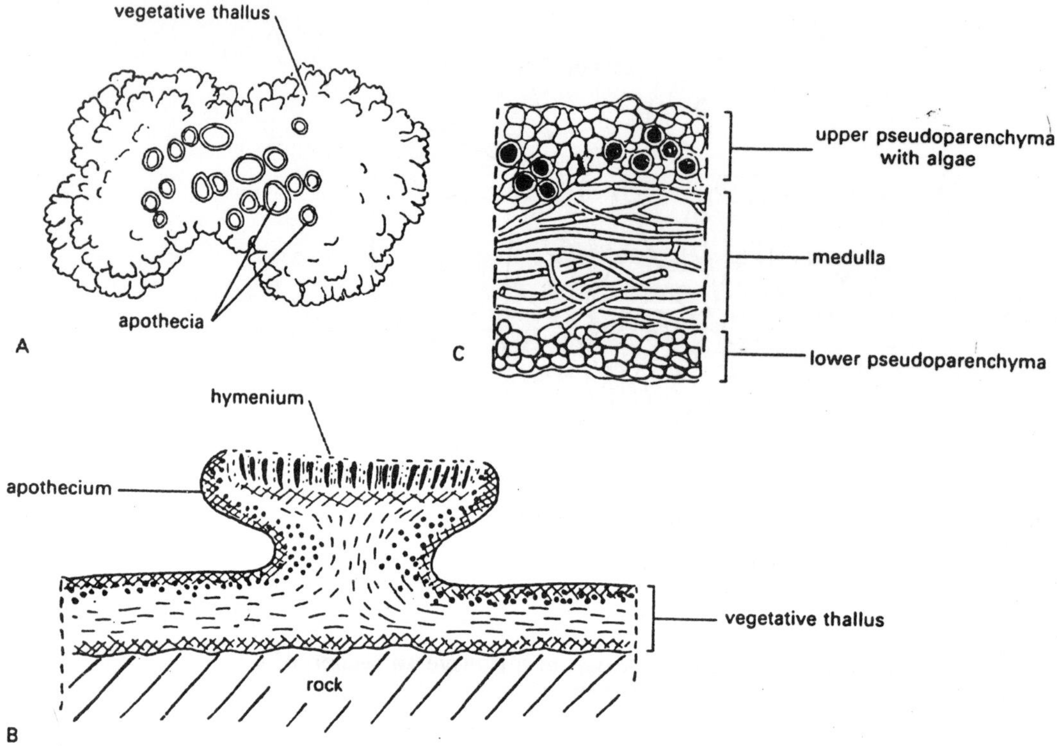

Fig. 11.5. *Xanthoria parietina*. A, thallus seen from above showing apothecia on its surface (life-size). B, vertical section of thallus passing through an apothecium; asci in hymenium shown black; the dots in the thallus are algal cells × 50. C, part of thallus showing nine algal cells in upper pseudoparenchyma (× 200).

Each lichen is named as if it were a single organism, i.e. it is given a Latin binomial.

In the lichen the fungal partner is nearly always an Ascomycotina belonging either to the Discomycetes or Pyrenomycetes or Loculoascomycetes. The autotrophic partner is either a blue–green cyanobacterium or a simple green alga. Both the latter are usually capable of an independent existence, whilst the fungal partner depends on its autotrophic associate.

The structure and mode of life of a lichen may be illustrated by considering a very common example, *Xanthoria parietina* (Fig. 11.5), an orange–yellow species found in great abundance on maritime rocks above high-water level and also on stone walls and roofs of farm buildings. The thallus forms a discoid crust, up to several centimetres in diameter, firmly attached to the rock except for its slightly upturned edge. In vertical section the vegetative thallus is seen to consist of four zones: an upper pseudoparenchymatous cortex of thick-walled, heavily gelatinized hyphae firmly cemented together; an 'algal' layer where the cells of the autotroph are surrounded by thin-walled, loosely packed hyphae; a medulla of loosely packed, thick-walled hyphae and a lowermost zone of pseudoparenchyma. Cells of the green alga *Trebouxia* occupy the algal layer usually occupying some 5–10% by volume of the thallus and varies from 5–10 μm in thickness. It is to be noted that *Xanthoria*, like the great majority of lichens, grows extremely slowly, increasing the radius of the colony by only about 2 mm a year. Although the thallus is so firmly attached to the substratum, it probably derives little of its inorganic needs from the rock. These seem to be met from rain and spray.

In *Xanthoria* sessile apothecia are formed on the upper surface of the thallus as tiny orange saucers (Fig. 11.5). In vertical section each is seen to have the same structure as the apothecium in an ordinary discomycete. Each ascus in the hymenium contains eight ascospores which are violently discharged. A new lichen thallus can develop only if an ascospore germinates alongside green cells of *Trebouxia*. However, in many species the reproductive process is not so haphazard. Many lichens produce powdery soredia over the whole surface or on a specialized part of the thallus. Thus in species of *Cladonia*, so common on dry heaths, the thallus is covered by a grey powder, each grain being a soredium consisting of a few algal cells wrapped around with fungal hyphae. The soredia are easily blown away and can grow to give new lichen thalli.

Lichens are outstanding colonists of such uninviting habitats as bare rock and the bark of trees. They show a considerable range of growth forms. *Xanthoria* and many others are encrusting, but some are endolithic with the thallus actually within the surface substance of the rock. Others are leaf-like and somewhat resemble thalloid liverworts. A well-known example is the dog's tooth lichen, *Peltigera canina*. A number are fruticose being like little shrubs. This is so in *Ramalina siliquosa* which commonly forms a grey turf on exposed sea cliffs. In another familiar fruticose species, *Usnea barbata*, the lichen thalli hang in grey tresses from branches of trees particularly in damper parts of the country.

It has been shown that the fungus lives at the expense of organic substances photosynthesized by its autotrophic partner. The fungus induces its partner to release a simple carbohydrate, such as glucose, or a polyhydric alcohol, such as ribitol or sorbitol, and uses this as its sole carbon source, and perhaps more importantly, thereby ensuring a low osmotic potential in the hyphae. This may be significant in surviving drying and rewetting. In those lichens, such as *Peltigera*, containing a cyanobacterium, the fungus receives both carbon- and nitrogen-containing substances from its autotroph because of the nitrogen-fixing ability of the cyanobacterium. It is difficult to demonstrate that the autotroph receives any benefit from imprisonment in the lichen thallus. There is no unequivocal evidence that any compound moves from fungus to alga nor does the fungus stimulate the growth of the autotroph as it does in mycorrhizal associations. However the habitat range of the autotroph is greatly extended. They grow where neither partner can alone.

Lichens are valuable indicators of air pollution since most species are extremely sensitive to sulphur dioxide in the air. Close to centres of heavy industry the bark of trees is almost bare of lichens. A few miles away only very resistant species, such as *Lecanora conizaeoides* can grow. A distance of 16–32 km may be required for a normal lichen flora to develop on trees (Fig.

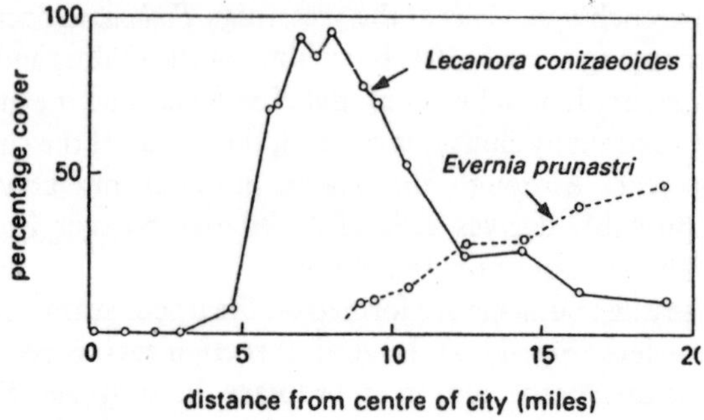

Fig. 11.6. Percentage cover by lichens on tree-trunks of ash (*Fraxinus*) travelling westwards from the City of Newcastle in northern England. (Based on graphs by Gilbert)

11.6). Lichens vary in their tolerance to sulphur dioxide and around such centres and cities concentric patterns of exclusion of particular species occur. These patterns are related to the mean winter sulphur dioxide levels in the atmosphere. A qualitative scale for the estimation of these mean levels has been devised using epiphytic lichens growing on tree bark. *L. conizaeoides* occurs even in many city centres. It is the most tolerant of all lichens to high sulphur dioxide levels. This can be explained in part by its relative unwettability so that the sulphur dioxide cannot penetrate. It is also a very poor competitor with other lichens and it seems probable that the elimination of its competitors by levels of sulphur dioxide which it can tolerate leads to its abundance in polluted areas. Lichens are equally valuable as indicators of atmospheric amelioration. *Hypogymnia physodes* is showing tentative re-establishment in a number of industrial areas in West Yorkshire. This is no doubt a result of Clean Air Legislation.

Seaweeds, such as fucoids, kelps and red algae, frequently support fungi as parasites or saprotrophs. *Mycosphaerella ascophylli* is always associated with *Ascophyllum nodosum* or *Pelvetia canaliculata*. Every plant is infected early in its life even before they are 5.0 mm in length. The fungus is systemic with a fine mycelium ramifying throughout the intercellular mucilage of the host, although it causes no damage. These seaweeds produce their sexual organs in conceptacles on specialized branches. In spring these bear minute, black pseudothecia of the fungus intermixed with them. There is no evidence of cell death or damage in the host and this has led to the suggestion that the two species of fucoids are lichenized. Certainly the fungus and algae never occur separately. They are both interdependent. Uninfected young algal plants do not survive beyond a particular stage. Both algae occur in the upper intertidal zone and one suggestion has been made that, like lichens, the fungus and alga in association are more tolerant of desiccation.

Mutualistic symbiotic associations also occur between fungi and insects. The ability of many insects to exploit potential food resources to their best advantage or even, perhaps, at all may depend upon an association with a fungus.

There are a number of ectosymbiotic associations where the fungus lives outside the insect and it either creates a favourable environment for the insect, for example it may shelter or protect the insect from the physical factors of the environment or from predators, or it becomes the food supply of the insect or it renders food digestible by providing enzymes for the insect.

Termites are best known as destroyers of wood. Some possess mutualistic bacteria and protozoa in their hindgut which hydrolyse ingested cellulose for the termite. The majority of termites lack these endosymbionts. They construct steeple-like mounds in tropical savannahs and maintain fungus gardens or 'combs' in these. The workers collect and eat plant debris, mainly wood, and use their faecal pellets to make the combs in chambers in the mounds. Each comb becomes colonized and permeated by the mycelium of a fungus. The majority of the fungi found in such combs belong to the Hymenomycete Agaricales in the saprotrophic genus *Termitomyces*. All species of *Termitomyces* are known to have obligate associations with termites. Most termites grow only one fungus in their combs. Small spheres or nodules, 0.5–2.0 mm in diameter, develop over the surface. These are the asexual or conidial stage of the fungus and they, and the mycelium, form the most important part of the diet of the termites.

In these fungus-growing termites woody tissues are finely chewed and the cellulose in the plant cell-walls is digested in the midgut. The termites starve if fed only on cellulosic material without access to the fungus. Cellulase is a complex of enzymes involved in the hydrolysis of cellulose. The entire enzyme complex can be found in the midgut of the termites but the salivary glands and the midgut epithelium only produce part of the complex. The important component which initially cleaves the cellulose into linear glucan chains which is vital for cellulolysis is not produced by the termites. They feed on the fungus and acquire this component when they digest the fungus, especially the nodules. From such evidence the plausible hypothesis has been put forward that many of the lytic enzymes involved in plant cell-wall degradation in the guts of arthropod detritivores may be fungal enzymes acquired while feeding on fungi or on substrates colonized by fungi.

Basidiocarps of *Termitomyces* are found early in the rainy seasons but only from mounds which have been abandoned by the termites. Presumably in the mounds the fungus is so heavily grazed that it never builds up sufficient reserves to reproduce. The termites obviously depend upon the fungus but the fungus also benefits by being 'cultured' in a favourable and equable microclimate as well as a nutrient-rich one.

Fungi and human affairs 12

Apart from the attack by so many parasitic species on our principal food plants, especially cereals and potatoes, fungi have a profound influence on human affairs. A few are themselves interesting foods; many are of considerable medical significance, and some produce intoxicants. There are even a few that, through their hallucinogenic properties, may have played a part in the origins of religious experience.

The cultivation of mushrooms in Europe has been going on for nearly 300 years. It began in the underground caves of Paris in the time of Louis XIV and, indeed, these are still in use for the same purpose. Now there is a flourishing worldwide mushroom industry with a total production exceeding 1.0 million tonnes per annum. Britain is the fourth largest producer being surpassed only by the USA, France and Taiwan. In Britain about 70 000 tonnes of mushrooms, valued at £200m at retail prices, are produced annually.

The species grown is *Agaricus bisporus*. It has not proved possible to bring the field mushroom (*A. campestris*), which has a better flavour, into effective cultivation. Mushroom production is a highly organized business. It is perhaps more of a factory than a farming operation. In more technological terms it is the application of solid substrate fermentation techniques to produce a valuable culinary commodity from the ligno-cellulose wastes of agriculture. The cultivation of *A. bisporus* involves two quite distinct sets of processes – the preparation by composting of a suitable growth substrate and its utilization for mycelial growth and reproduction of the fungus (Fig. 12.1).

Composting itself occurs in two phases. Traditionally the starting point has been stable manures but a mixture of wheat straw and manure or other nitrogen rich material (activator) is now used. The compost is first stacked outside. The stacks (windrows) are mechanically turned to mix and are wetted at 2–4 day intervals for 7–14 days. The temperature in this phase may rise to 75°C. The turning process breaks down large gradients of temperature, oxygen, carbon dioxide and moisture as they build up and ensures that the population of micro-organisms is evenly distributed. This brings about a more uniform degradation and consequently a more homogeneous product. The heat comes from the microbial metabolism in the compost and because of the insulatory properties of the straw is retained within it. The wetting replaces

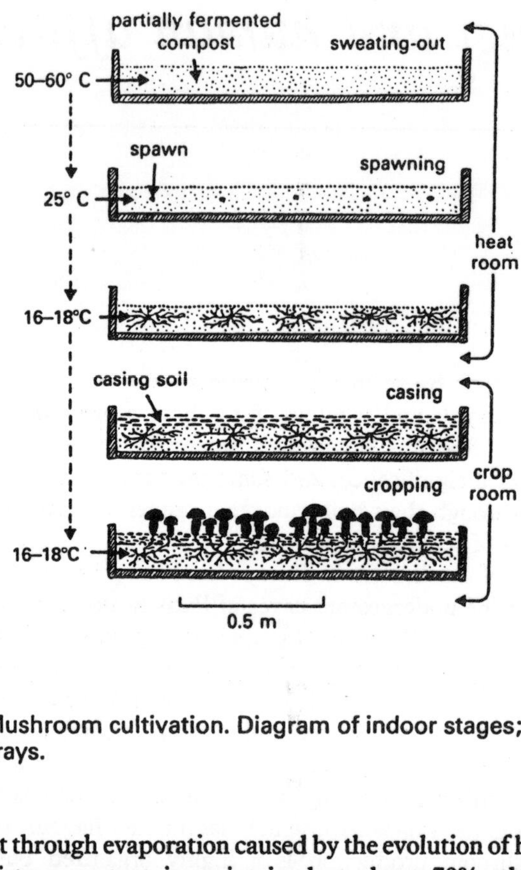

Fig. 12.1. Mushroom cultivation. Diagram of indoor stages; the compost is in wooden trays.

any water lost through evaporation caused by the evolution of heat and ensures that the moisture content is maintained at about 70%, the optimum for microbial growth. The addition of manure or other nitrogen rich activators ensures that sufficient nitrogen is incorporated for the subsequent growth of the mushroom. During this phase there is a build-up of a specialist group of thermophilic actinomycetes, bacteria and fungi. These become important in the second phase of composting when the stacks are broken up and the compost filled into wooden trays 1200 × 750 × 120 mm. These are placed indoors in controlled environmental conditions for the pasteurization or 'peak heat' phase. The temperature is maintained at 55–60°C for 4–8 days. There is no mixing. During this period the thermophiles utilize all the soluble carbon and nitrogen sources which were present initially. Any ammonia, which is toxic to *A. bisporus*, is removed by being incorporated into microbial protein. Finally the compost may be treated with steam to achieve a temperature of 70°C for 1–2 hours to kill off any potential pests and disease causing organisms. The temperature is allowed to fall and the compost is in a condition for 'spawning'. The substrate, the compost, now favours the growth of a slow-growing agaric fungus against a wide range of competitors. 'Spawn', the inoculum, is prepared by the large scale production of axenic cultures of the mycelium of the fungus on sterile cereal grains, such as those of millet and rye. This is the only part of the whole cultivation which takes place under sterile conditions.

The compost is then thoroughly mixed with the 'spawn' on a conveyor belt line of mixing machines and is pressed back into the trays. The pressing increases the bulk density. This cuts down loss of moisture and favours colonization by *Agaricus*. The trays are then stacked in the cropping rooms at a controlled temperature of 25°C, the optimum growth temperature for the fungus. Low ventilation is used. This increases the ambient carbon dioxide concentration which appears to favour the growth of the fungus. Over the next 10–14 days the 'spawn runs'. *Agaricus* rapidly colonizes the compost. It grows best on slowly hydrolysable organic carbon and nitrogen sources. Conditions in the compost are conducive for *Agaricus* to excrete extracellular hydrolytic and oxidative enzymes into the compost to degrade hemicellulose, cellulose, lignin and the dead remains, especially the protein, of the micro-organisms which preceded it in the compost. The microbial biomass formed in the compost by the end of the second phase of composting accounts for about 10% of the total carbon nutrition of *Agaricus*. It may also supply nitrogen, phosphorus and other minerals in a suitable form for absorption.

About two weeks after spawning the mycelium should have penetrated throughout the compost. 'Casing' then occurs. The compost is covered with a layer, 10–30 mm deep, of a peat chalk mixture. This is an essential step in the commercial production of basidiocarps, but its physiological basis is not really known. It certainly requires the presence of living micro-organisms in the casing material.

After casing, the temperature in the cropping rooms is lowered and maintained at 16–18°C and the ventilation increased. As in other agarics which have been studied critically, the optimum temperature for reproduction is lower than that for vegetative growth. Ventilation is increased to lower ambient carbon dioxide levels. Levels are not allowed to rise above 0.5%. This is important not only in initiating basidiocarp production but higher levels may cause deformation in the shape of the mushroom. Water may also be added to maintain the moisture content of the compost and pesticides and fungicides may be applied to control pests and fungal diseases of the mushroom.

Basidiocarps appear after a further 2–3 weeks in a series of flushes at 7–10 day intervals. After four to five flushes the compost is 'spent' and no longer commercially productive. 'Cook-out' occurs. The cropping rooms are steam treated to kill off and prevent carry over of pests and pathogens to the next cycle. The total productive cycle takes 75–100 days.

Throughout the process of mushroom production, the mycelium is competing with other micro-organisms in the compost and the procedures adopted by the grower are designed to make the compost most conducive to the growth of *Agaricus* to ensure that it eventually dominates this modified microbial succession. As in any cultivation system, pests and disease can cause serious problems if left uncontrolled. Sciarid and phorid flies and nematodes may cause damage by feeding on the mycelium and disfiguring the mushrooms as they develop. Parasites include bacteria, fungi and viruses. A number of fungi such as species of *Mycogone* and *Verticillium* produce severe abnormalities of the basidiocarps rendering them unmarketable. *M. perniciosa* has no, or very little, effect on the mycelium of *Agaricus*. Disease symptom expression is related to the stage of basidiocarp development at the time at which infection occurs. If infection occurs early, there is no development of the stalk, cap and gills. The basidiocarps are totally deformed and they develop into spherical masses of undifferentiated tissue. Infection of well-developed basidiocarps results in the pathogen growing over sectors of the gills and the cap above them. The gills hypertrophy giving them a sinuous appearance. These pathogens have been controlled using fungicides with low toxicity to the mushroom itself but problems have been encountered with the evolution of tolerance to these. Control has reverted to good hygiene to avoid infection. Saprotrophic cellulolytic and lignolytic fungi, such as species of *Coprinus*, the 'ink-caps', may invade the compost and sometimes outcompete *Agaricus*.

In South-east Asia, the padi-straw mushroom, *Volvariella volvacea*, is extensively grown on raised beds of damp rice straw. The beds consist of four to seven layers of bundles of straw, some 400 mm in length and 100 mm in diameter, with a 200 mm layer of loose straw on top. They are built under the shade of trees or creepers and covered with a sheet of polythene supported on a bamboo framework. This maintains a fairly equitable moisture and temperature regime. A simple method of preparing spawn is generally used. Young mushrooms are chopped up with a knife, mixed with rice husk and husk ash and kept in a clean container overnight to encourage mycelial growth. Balls of spawn, about 25 mm in diameter are pushed 50 mm into the straw bundles at 100 mm intervals as the bed is made. Cropping occurs 10–14 days after spawning. Cultivation of this agaric is not so highly organized as is that of shiitake or of the ordinary mushroom, and is merely a sideline of other farming activities. Nevertheless yields may be high. A single layer bed may yield as much as 7 kg of closed or 17 kg of open or 'blossom' mushrooms.

In Hong Kong, more readily available residual cotton waste has completely replaced rice straw as a substrate for the padi-straw mushroom and it is grown entirely indoors. It not only produces a higher yield but basidiocarps develop earlier. Cropping occurs 8–10 days after spawning. The cotton wastes are usually mixed with small quantities of rice bran and calcium carbonate. A number of other fungi are grown on a smaller commercial scale in South-east Asia. These include the Abalone mushroom, *Pleurotus cystidiosus*, several other *Pleurotus* spp., *Auricularia* spp. and the golden mushroom, *Flammulina velutipes*. Most of these grow and crop well on a mixture of sawdust, cotton waste, rice bran and calcium carbonate, packed in 1 kg quantities in plastic bags. The latter are removed as basidiocarp primordia appear.

In Japan another agaric is cultivated. This is the shiitake, *Lentinus edodes*, a wood-rotting species. Its large-scale production is an industry as highly organized as mushroom growing in Britain and the USA. The fungus is grown on freshly cut logs of oak 50–150 mm thick and over 1 m long. The spawn is supplied to growers as a pure culture on bits of wood, and these are inserted into holes drilled into the logs. The mycelium then grows in the wood. After about six months, when reproduction is ready to occur, the logs are placed in a damp environment at a relatively low temperature which encourages basidiocarp formation.

It is interesting to note that the only three agarics now widely grown for sale have all been in cultivation for hundreds of years.

Another fungus greatly prized as a luxury food is the Perigord truffle, *Tuber melanosporum*. This subterranean fungus grows in association with the roots of trees, particularly oak. In southern France and northern Italy, special plantations are established in which, as the result of appropriate cultivation on suitable soil, truffles develop by natural dissemination. This is a far cry from the controlled growth of mushrooms. Since truffles are hidden below ground, their collection presents a major difficulty. Dogs and pigs have to be trained to locate and dig them up. When ripe, truffles emit a powerful odour.

Many other fungi are collected for the pot from woods and fields. However, the British are not great fungus eaters, and usually only the cultivated mushroom is sold on vegetable stalls. In most other parts of Europe, however, many kinds of toadstool, duly authenticated, are on sale in markets, including boletes (*Boletus* spp.) and the chanterelle (*Cantharellus cibarius*).

The chief factor militating against the eating of toadstools is justifiable fear. Although most are harmless and many are good to eat, a few are deadly poisonous. The majority of deaths result from eating the death-cap (*Amanita phalloides*). This is a common woodland mycorrhizal species, about the same size as a field mushroom, with a greenish cap and white gills. There is a cup-like volva around the base of the stipe in addition to a ring higher up. Quite a small piece of a basidiocarp can cause death. Although the clinical picture of poisoning by this fungus is well-known, there is some uncertainty about what substances are the cause of death. There are two strong poisons in *A. phalloides*, phalloidin and amanitin, both cyclopeptides. The evidence now suggests that it is amanitin which is the killer.

The fly agaric (*Amanita muscaria*), with its warning red cap ornamented with white spots, is poisonous but is unlikely to cause death. The poison (muscarin) in that fungus is quite different from phalloidin and amanitin.

There are other fungi, besides toadstools, that cause poisoning. An outstanding example is *Claviceps purpurea*. The black sclerotia, ergots, of this species replace some of the ovaries in infected grasses and cereals especially rye (Fig. 10.1A). At harvest ergots may incidentally be gathered with the grain. If this is then fed to animals or milled for consumption, severe poisoning may result if enough ergot is present. In the middle ages, when certain European peoples, particularly around the Baltic, ate much rye bread, there were often outbreaks of ergotism leading to many deaths. The results of poisoning are convulsions and toxic spasms of the limbs and the gangrenous rotting of extremities, such as hands and feet, following constriction of the blood capillaries. The poisoning produces a burning feeling in the limbs and this has led to ergotism being known as St Anthony's Fire. The poisonous alkaloids, all with a chemical structure based on lysergic acid, that can be extracted from ergots have an important use in obstetrics. Cautiously administered, they restrict haemorrhage after child-birth. In this connection there is a great demand, and to supply this rye is deliberately infected with *Claviceps* to provide crops of ergot.

Many saprotrophic fungi commonly found growing on decaying plant matter are important because of their ability also to grow on food, animal feeds or the raw materials used in the manufacture of these. They are usually referred to as 'moulds' and are taxonomically very diverse. Some of the most common are from the genera *Aspergillus*, *Fusarium* and *Penicillium*. As they grow on these substrates, they produce a wide range of secondary metabolites. Some of these are toxic to man and his domestic animals, causing illness and even death. They are called mycotoxins. Illnesses caused by the direct consumption of such mycotoxins are called primary mycotoxicoses but the toxins may be passed on in the food chain into animal products, such as milk and meat, causing secondary mycotoxicoses. Not only are these mycotoxins chemically very diverse, they also cause a wide variety of effects. For example, some are carcinogenic, others cause extensive necrosis and haemorrhage of the liver or degeneration of the proximal tubules in the kidney.

The best known, or most infamous, of these mycotoxins are the aflatoxins produced by some strains of *Aspergillus flavus*. It has been known to cause many deaths from acute liver disease in young turkeys and trout which have been fed with groundnut meal made mouldy by *A. flavus*. Aflatoxins cause the cells of the liver to break down and also block the hepatic veins. Death results from loss of liver function.

In many animals aflatoxins are known to be carcinogenic, producing tumours in the liver. During the complex reactions accompanying the metabolic detoxification of one of these, aflatoxin B_1, in the liver, a highly reactive epoxide of the aflatoxin is generated which binds with DNA, disrupting transcription, and with RNA and proteins, reducing enzyme production and functioning. There is now sound epidemiological evidence that this form of cancer, common among human populations in Third World countries, such as East Africa, the Philippines and Thailand, but rare in developed countries, results from exposure to aflatoxins. This is brought about by eating food such as groundnuts, beans and grain made mouldy by the growth of *A. flavus* under warm, damp tropical and sub-tropical conditions.

A number of fungi growing in a watery medium of suitable composition produce compounds of great importance to man. Outstanding in this connection is the production by yeast of ethanol. Not only are wines, beer and spirits of great significance but alcohol is also used extensively in industrial processes.

The species most concerned is the brewer's or baker's yeast, *Saccharomyces cerevisiae*. In a well-aerated glucose solution yeast grows vigorously and respires normally, the sugar being broken down according to the overall equation:

$$C_6H_{12}O_6 + 6O_2 = 6CO_2 + 6H_2O + 2880 \text{ kJ}$$

In the absence, or near absence, of oxygen growth ceases and respiration becomes of a type known as fermentation in which much less energy is released for each molecule of glucose consumed. The process is summarized by the equation:

$$\underset{\text{glucose}}{C_6H_{12}O_6} = \underset{\text{ethanol}}{2C_2H_5OH} + 2CO_2 + 210\ \text{kJ}$$

However, many intermediate steps are involved.

As an example of the commercial importance of yeast, beer, produced from barley, may be considered. The food reserve of barley is starch which yeast cannot hydrolyse since it lacks the necessary enzyme, amylase. The initial process in making beer is malting. In this the barley grains are allowed to germinate to the stage when the hydrolysing enzymes have just started to be active. Germination is then stopped by an increase in temperature which, however, is not enough to destroy the enzymes. The grain, with its food reserves and active enzymes, is next ground up as the malt. This is prepared by the maltster and supplied to the brewer. His initial process is mashing. In this the malt is mixed with warm water with the result that the amylase of the barley grains converts all the starch into sugar. At the appropriate stage the enzymic process is irreversibly stopped by a sufficient rise in the temperature. The clear fluid is run off, hops are added and the whole is boiled. The liquid (wort) is cooled and transferred to the fermentation tank where yeast is added and the wort is converted into beer by fermentation. The substances extracted from the hops not only give the beer its characteristic flavour, but also prevent the activity of certain bacteria which might transform some of the alcohol to acetic acid and lead to souring.

Of the two products of fermentation of sugar by yeast, one, alcohol, interests the brewer, the other, carbon dioxide, is the concern of the baker. Yeast in his dough ensures a uniform production of gas leading to a well-risen loaf. The baker now normally uses compressed yeast specially prepared for him and no longer relies on supplies from the brewer.

Shoyu, or soy sauce, is produced in a two-stage fermentation by growing the filamentous fungi *Aspergillus oryzae* and *A. soyae* on a mixture of soybeans. These are cooked with steam under pressure and incubated with the fungi and roasted wheat grains for 48–72 hours at 30°C at a moisture content of about 40% to form koji. This short, first-stage fermentation produces the necessary enzymes to hydrolyse the proteins and starch in the seeds. The koji is mixed with salt water to make a mash, moromi, and after inoculation with yeasts and lactobacilli is transferred to deep fermentation tanks for 4–8 months. The filamentous fungi in the koji use about 20% of the starch in the seeds in their growth and their amylases convert the remainder to simple sugars and their proteases hydrolyse the proteins in the seeds to amino-acids. Over half of the sugars are then fermented to lactic acid and ethanol by the lactobacilli and the yeasts respectively. The liquid is pressed from the moromi and pasteurized to inactivate the enzymes and kill the micro-organisms. A good quality shoyu contains 1.5–2.0% nitrogen, over half of which is in the form of free amino-acids, 3.0–5.0% reducing sugars, 2.0–2.5% ethanol, 1.0–1.5% glycerol, a variety of vitamins and just under 20% sodium chloride at a pH of 4.7–4.8. Miso is produced in a similar manner to shoyu, with the finished product in the form of a paste. Typically it is used as a soup base but also to add flavour to a wide variety of foods.

There is a multitude of such two-stage fermentations used to produce highly flavoured, major nutrient rich foods in the East; mixed culture inocula or starters are commonplace in local markets. For example, ragi is a mixture of spores of *Mucor* and *Rhizopus* with yeasts and bacteria in a rice flour base. Starters containing very similar micro-organisms are used in such diverse countries as China, Indonesia, Nepal and the Philippines, to saccharify cassava, rice or sorghum and to make alcoholic beverages or fermented foods.

The 'Amylo' process is used worldwide in the saccharification of cereal starch to be converted to alcohol. Sake, the national drink of Japan, uses rice koji and sake yeast, *Saccharomyces sake*. A drawback of such processes is that 30–40% of the total energy input into ethanol production is in cooking the starch before fermentation. Much interest is now centred on the amylases of *Rhizopus* which can hydrolyse raw starch. This not only saves energy but recovery yield of sugars is about 60%, compared with only about 20% with cooked starch and *Aspergillus* amylases.

A number of these oriental fermentations are now becoming popular in the West. Tempeh, an Indonesian fermented food, is one of these. It is made from cooked, de-hulled soybeans. The beans are packed in a flat container, inoculated with spores of *Rhizopus oligosporus* and incubated at 30°C for 24 hours. The beans become covered by the mycelium of *Rhizopus* to form a white solid cake. This is cut into slices, sprinkled with salt and fried in oil. The fermentation relies upon the production of lipolytic and proteolytic enzymes by the fungus. This not only enhances the digestibility of the beans but it also makes them more palatable by increasing their vitamin content. In Indonesia it is used as meat substitute. In the USA it is popular as a non-animal protein in the form of tempeh burgers replacing the familiar beefburger. The current retail value there is over US $2m.

The concept of using filamentous fungi as an alternative protein food for human consumption was developed in the 1960s by Rank, Hovis and McDougall. They realized that the development of single-cell protein from yeasts and bacteria for use as animal feed to boost meat production was not an economic proposition. At this time there was a real concern that in the foreseeable future there would be a global shortage of traditional protein foods. They screened some 3000 isolates of filamentous fungi and chose *Fusarium graminearum*. When grown on a glucose–ammonia–biotin–mineral salts medium, its mycelium had one of the highest protein yields (over 42% w/w). For production on the commercial scale to enhance productivity and to produce sufficient volume, the fungus has to be maintained in a perpetual exponential growth phase. This is achieved in a 40 m^3 volume air-lift fermenter in which the volume of culture is kept constant by removal at an identical rate to which fresh culture medium is supplied (Fig. 12.2). The product registered as 'Quorn' myco-protein can replace 75% of the protein in our diet. This is comparable with cow's milk. When supplemented with 0.2% methionine, it can provide a total replacement. Single-cell yeast protein can only replace about 10% of our protein requirements. Perhaps the most

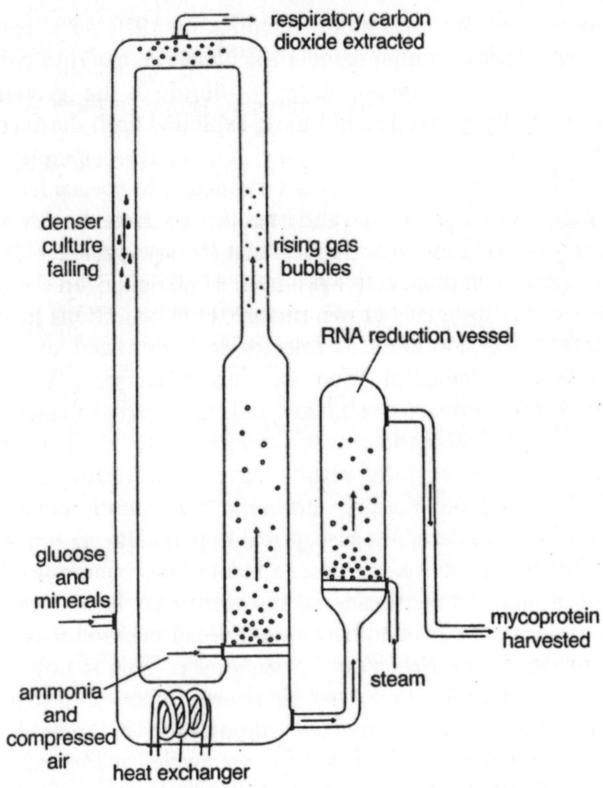

Fig. 12.2. Sectional diagram of an air-lift fermenter (after Trinci). The fermenter has no moving parts. Continuous circulation is maintained using the difference in specific gravity of the rising aerated culture and the falling air-depleted culture. The RNA content of the fungus has to be reduced to meet World Health Organization guidelines.

important features of 'Quorn' are its fibrous texture, its chewiness and its succulence. It closely resembles meat. Now, in the 1990s, the predicted world shortage of animal protein has not yet materialized but 'Quorn', in the form of savoury pies, casseroles, etc., is helping to satisfy current dietary trends and retail sales are over £12m per annum in the UK. It provides a healthy low calorie food (334 kJ $100g^{-1}$) of high quality protein, high in dietary fibre (5.1 g $100g^{-1}$), low in saturated fats and with no cholesterol. It also contains the correct proportions of the eight amino-acids required by man.

One of the most important products of filamentous fungi is citric acid, great quantities of which are needed for the soft-drink industry and for medicinal purposes. Formerly the supply came from citrus fruits, but now it is nearly all of fungal origin. The organisms used are species of *Aspergillus*, mostly *A. niger*. The fungus is usually grown in shallow pans containing a sucrose solution and the necessary inorganic salts. For every gram of sugar consumed about half that weight of citric acid may be liberated into the culture medium.

Perhaps the most spectacular by-product of mould activity is penicillin.

This was discovered by Alexander Fleming in 1929 as a result of his observation that a chance contaminant (*Penicillium notatum*) in a culture of the bacterium *Staphylococcus aureus* strongly inhibited the growth of that pathogen. The discovery was not, however, exploited until the Second World War, and it then revolutionized medicine. The active principle, penicillin, was isolated and its chemical structure determined. The special features of this antibiotic are its low toxicity to man and its ability to arrest the development of gram-positive bacteria. Later it was found that *P. chrysogenum*, closely related to *P. notatum*, is an even more active producer of penicillin. In the production of the antibiotic, the fungus is grown under sterile conditions in submerged cultures in large tanks continuously injected with sterilized air.

The discovery of penicillin stimulated an extensive search for other antibiotics produced by micro-organisms; and this greatly advanced the study of mould fungi in particular and of mycology generally. Having regard to the enormous efforts involved, the results have been rather disappointing. However, one further outstanding antibiotic was found, namely streptomycin, which affects bacteria not susceptible to penicillin. It is not produced by a fungus, but by *Streptomyces griseus*, a filamentous bacterium. Nevertheless, many antibiotics, both antibacterial and antifungal, have been discovered and some of these are of considerable value. Special mention may be made of griseofulvin produced by *Penicillium griseofulvum*. This is now used as an effective drug, taken orally, to control ringworm infection in man.

While considering the commercial importance of members of the Aspergillaceae, reference may be made to two species of *Penicillium* used in the production of certain cheeses.

P. camemberti is used in the ripening of soft cheeses such as Camembert and Brie. The fungus develops on the outside of the cheese and only grows a short distance into it, but this is enough to give the desired texture and flavour. Usually there is no need to inoculate the cheese with *P. camemberti* for, in a factory devoted to its production, its spores are sufficiently prevalent in the air.

In hard, blue-vein cheeses such as Roquefort, Stilton and Gorganzola, the fungus used in ripening is *Penicillium roqueforti*. In these cheeses the fungus is in nearly pure culture and other species of *Penicillium* cannot compete. The special physiological features which determine its dominance within the cheese are its ability to grow at very low concentrations of oxygen and to endure relatively high concentrations of carbon dioxide. In the manufacture of blue-vein cheeses, the fungus is introduced into the curd as crumbs of bread on which the mould has been grown.

Fungi are of great importance in relation to timber. Standing trees, sawn and seasoning wood, timber in houses and boats are all liable to deterioration resulting from the growth of fungi. The rotting of standing trees has been considered in an earlier chapter (p. 162). Cut timber is also prone to attack, but usually the fungi most involved are different from those that infect unfelled trees. In the colonization of wood the major factor is water content. If this is less than 20%, wood is generally safe from fungal attack. However, a local damp region may allow a fungus to gain a foothold, and once growth starts, respiration, involving the breakdown of the cellulose of the wood, produces water which may permit further extension of the mycelium.

Lentinus lepideus is among the commonest of the species responsible for the decay of telegraph poles, wooden paving blocks and pit props in mines. Another abundant species is *Peniophora gigantea* with irregular resupinate basidiocarps. It is extremely common on coniferous logs and soft-wood planks in storage.

The blue-stain fungi cause timber to lose marketable value without, however, impairing its mechanical properties. Mainly involved are species of *Ceratocystis*, a genus of Ascomycotina, which have perithecia with long necks and also reproduce by various kinds of conidia. The grey–blue discolouration of the wood is due to the pigmented hyphae which grow in the wood parenchyma and rays of the sap-wood. Coniferous wood is especially liable to blue-stain.

In the temperate regions of Europe, enormous damage is caused to timber in houses by 'dry-rot', an unfortunate term since this sort of decay is always connected with extreme dampness somewhere. Most kinds of wood are affected although oak in houses is resistant to rot. The fungus concerned is *Serpula lacrymans*, a Basidiomycotina belonging to Aphyllophorales.

Dry-rot, in which the wood becomes soft and broken into small cubical portions, is especially prevalent in old houses where the ground floor is constructed of planks with a cavity below, which if ventilation is poor, may easily become damp. Again timber in cellars is very liable to rot. The trouble begins if, in some part of a house, the woodwork becomes damp so that its water content rises well above the critical 20% level. The seat of the trouble is often hidden, although the presence of dry-rot is usually betrayed by a characteristic musty smell and sometimes by the accumulation of a rusty deposit of basidiospores on polished surfaces. The basidiocarps develop, mostly well out of sight, on beams and joists in damp cellars or under floor-boards, as irregular, honey-coloured structures with a downward-facing hymenial surface from which the brown basidiospores are shed in enormous numbers (Plate 2a). The basidiocarps often weep drops of water, a feature responsible for the specific epithet (*lacrymans*). A special feature of *Serpula lacrymans* is the development of hyphal strands, rather like the rhizomorphs of *Armillaria* (p. 162) but not so highly organized, which may be several millimetres thick. These, arising from an established mycelium of the fungus, can grow for distances of up to several metres over non-nutrient material, such as brickwork, and then on contact with wood, initiate a new centre of rotting. They arise as one or more wide leading hyphae, the strand initials. The branching pattern of these changes and instead of their branches growing out at an acute angle by thigmotropic growth they ensheath the initials. The branches, 'tendril hyphae', coil around the initials. Lateral fusions occur between the branches. The large, now central, initials and their larger branches may have their septa broken down and they become open tubes or 'vessel hyphae' and serve to conduct nutrients and water. Some of the laterals grow intrusively through the strand and develop into narrow, thick-walled 'fibres' to give a degree of support. The whole becomes stuck together by adhesive extrahyphal material. Such wood decay fungi as *S. lacrymans* may find that the food base that they are about to colonize to be poor in readily available carbon and nitrogen sources. They depend upon the translocation from behind of sufficient energy sources to initiate the process of cellulose degradation so that they can become established.

At the start of the nineteenth century ships of the Royal Navy were, perhaps, more at risk from attack by fungi than from men-of-war of Napoleon's fleet. Things came to a head with the case of the 'Queen Charlotte'. This, a fine warship of 110 guns, was launched in 1810, but rotted so quickly that extensive rebuilding was needed even before she could be commissioned for service at sea. The naval authorities in England became really alarmed and engaged James Sowerby, a noted botanical artist much interested in fungi, to study the problem. He examined the 'Queen Charlotte' and other warships, and in due course produced a long report illustrated with coloured plates of the species he had found in the ships. He also gave good advice emphasizing the need to use properly seasoned timber in ship-building if wood-rot was to be minimized.

Besides the rotting of wood, there are other kinds of biodeterioration of considerable economic importance in which fungi play a major role.

Paper is a major cellulosic product especially liable to degradation by fungi. Wood pulp may be attacked by a variety of Basidiomycotina, which weaken the fibre strength well before the paper is made. Paper itself is subject to attack by a number of cellulolytic Ascomycotina, particularly those in the genus *Chaetomium*. Members of that large genus have isolated perithecia beset with long hair-like hyphae (p. 64). *C. globosum* may be troublesome on newsprint if stored under conditions that are not dry enough and on wallpaper in damp rooms.

During the Second World War the American forces in the Pacific suffered great losses of clothing, tentage, optical and other equipment owing to the activity of moulds, mainly species of *Aspergillus* and *Penicillium*. This led to major studies in biodeterioration. Fungicides were found to help, but the only sure way to prevent fungal growth was to keep materials dry, often difficult in the humid tropics. Glass lenses in spectacles and optical instruments may be etched by fungi growing over their surfaces utilizing susceptible coatings and organic debris present on the glass. Etching is a result of organic acid production. Aluminium alloy in aircraft fuel tanks may become pitted by a variety of organic acids, such as citric and oxoglutaric acid, produced by *Cladosporium resinae* as it metabolizes hydrocarbons in the kerosene fuel. It is a soil-inhabiting fungus which can be selectively isolated from kerosene or creosote impregnated matchstalks placed in soil. Even paint surfaces are susceptible to fungal deterioration. *Aureobasidium pullulans* is frequently isolated from such films. It is a common phylloplane inhabitant and when subjected to desiccation or ultra-violet light it deposits melanins in its hyphal walls. It is these black pigments which disfigure the paint surfaces.

Even in the tinned-fruit industry fungi can be troublesome. Ascospores of *Byssochlamys fulva*, frequently present on fruit from orchards, can remain viable for 30 minutes in syrups at 86–88°C, and, therefore, are not always destroyed in the canning process. Since this fungus can grow with a low supply of oxygen, a mycelium may develop in the sealed cans and the fruit be spoilt.

Fungi are much more successful as saprotrophs and necrotrophic and biotrophic pathogens of seed plants than they are as pathogens of man. In man and other mammals, unlike plants, most diseases are caused by bacteria and viruses. Fungi are not well-adapted to living deep within animal tissues. The ability of many fungi to grow at human body temperatures conveys on these fungi the added potential to become pathogenic. This potential is more likely to be expressed when the host is stressed or predisposed in some way. Like plants, humans and other animals have defence mechanisms to prevent infection. These include physical barriers, such as intact mucocutaneous structures, toxic secretions and competitive and antagonistic yeasts and bacteria resident on surfaces, such as the skin and mucus membranes. Other mechanisms include humoral and cellular responses, such as cytokine and immunoglobulin production and the activity of neutrophils, monocytes and natural killer cells. One successful group of human pathogenic fungi is the dermatophytes or ringworm fungi. The commonest of these are species of *Microsporum* and *Trichophyton*, which invade skin, nails and hair and digest the keratin.

In head ringworm the hair follicles are colonized by the fungus with the result that the hairs break off near the base leaving a bald patch. The mycelium grows in the stump of each hair and this becomes surrounded by a sheath of minute arthroconidia which can spread the disease by direct or indirect contact. Identification of the pathogen depends on growing it in pure culture. It there produces the macroconidial state which is characteristic and allows precise identification to be made. In Britain two species of *Microsporum*, *M. audouini* and *M. canis*, often produce ringworm in children. The former is confined to human beings, but the latter also attacks cats and dogs. Although it can be grown in culture, *M. audouini* has no natural existence except as a parasite of man and it depends for survival absolutely on its host.

In athletes's foot, common among adolescents and young adults, the fungus, *Trichophyton interdigitale*, grows in the skin between the toes.

Candida albicans is another troublesome fungal pathogen. It occurs regularly in the human alimentary canal and is normally harmless. When the resistance of the body is lowered, however, it may attack the mucous membranes of the mouth causing 'thrush' especially in infants. In adults other parts of the body may be affected, particularly the vagina during pregnancy. Although in the human body the fungus exists only in the yeast condition, in culture on a suitable medium a mycelium is produced which, in addition to budding off yeast-like cells, forms terminal chlamydospores.

In parts of the Americas *Histoplasma capsulatum* is endemic and, whereas most infections are mild and not even noticed, it may cause an infrequent but serious, often fatal, disease. Inhaled spores may first develop and cause lesions in the mouth. Yeast-like colonies then develop and spread, especially to the lungs, but also to the liver, kidneys and spleen. As the yeast cells divide, the host cells multiply rapidly, such that infected tissues become disorganized, malfunctioning masses. Severe infections cause anaemia, fever and ulceration of tissues, with fatal results. The fungus also occurs as a saprotroph in soil enriched by the droppings of birds and bats. The disease develops particularly in those who have cleaned out old poultry houses or have visited caves where bats occur in great numbers.

Most other fungi which infect man are opportunists. They infect tissues well within the body. Such deep mycoses are almost always associated with abnormal host immunity occurring in patients with post-operative problems or after the use of broad spectrum antibiotics and steroid therapy. There is evidence of a marked increase in these opportunistic mycoses following the use of modern techniques used to treat very ill and debilitated patients. A similar situation exists in intravenal drug abusers. In such patients relatively harmless infections of *C. albicans* can cause a serious candidosis of the oesophagus, bronchi, lungs or other vital organs and become a chronic, debilitating disease. The spread of the HIV-virus associated with AIDS which removes all or most of the host's immunity system is leading to a marked increase in such deep mycoses. A major concern is cryptococcosis, caused by the Basidiomycotina yeast *Cryptococcus neoformans*, which lives, like *H. capsulatum*, saprotrophically in bird droppings, especially in those of pigeons. It may survive for up to 10 years in these. Levels as high as 5×10^7 per gram of droppings have been recorded. The fungus occurs as non-capsulated yeast cells, 1–12 μm in diameter. The longer the cells remain in the droppings or soil the smaller they become. Some 7–15% of the cryptococci extracted from dry pigeon droppings are in the range of 1.1–3.3 μm in diameter. Cells smaller than 2.0 μm, once airborne, can be deposited on the walls of the alveoli, where they start to grow by producing capsular material.

Infection arises from inhalation of such airborne spores which cause quiescent lesions in the lungs. In the tissues, these spores enlarge and possess a thick mucilaginous capsule, which prevents the normal phagocytosis of such foreign cells. These persist as a small primary focus of living fungus. Subsequent development depends upon the resistance and immunological response of the host. Spread from the respiratory tract foci occurs especially in patients with diminished monocyte and inflammatory responses and can result in systemic spread with lesions in almost any organ of the body. Infection of the central nervous system and brain are the worst complications. The commonest symptoms are those of chronic meningitis. With infection of the central nervous system symptoms range from a serious relapsing disease in apparently healthy people to an acute illness, often resulting in death, in severely predisposed patients such as those undergoing corticosteroid therapy or with very low levels of leucocytes. Although relatively few cases are reported each year from England and Wales, from 1984, the number has steadily increased and most have been associated with AIDS. In Zaire, cryptococcosis has been found in almost one-third of all recorded AIDS patients.

There is a further medical aspect of the fungi, for spores may act as allergens. The respiratory tract, consisting of the nasal system, trachea, bronchial tubes and alveoli of the lungs, is a kind of impactor trap (p. 123). Air is sucked in and the suspended spores are deposited in various parts, the biggest first with the smallest penetrating deepest. In the nasal regions the largest elements of the air-spora, particularly the pollen grains, tend to be impacted. The abundant pollen of grasses may induce hay-fever in susceptible persons. Fungal spores are generally much smaller and are likely to be deposited mainly in the trachea and bronchial tubes. Only the smallest spores, such as those of *Penicillium* and *Aspergillus*, reach the deepest recesses of the system.

Some types of asthma are caused by allergic reaction to fungal spores. Workers in glasshouses with tomatoes heavily attacked by *Cladosporium fulvum* can be strongly affected, those living in old houses where there is dry-rot may become asthmatic from inhaling spores of *Serpula lacrymans*, and teliospores released from smutted cereal crops may also be responsible for asthma amongst the harvesters.

The common thermotolerant mould, *Aspergillus fumigatus*, which grows well on self-heating materials, such as garden composts, can liberate vast numbers of conidia that are able to penetrate deeply into the respiratory system. In this case its conidia, and those of other aspergilli, such as *A. flavus*, may not only bring about an allergic reaction, but may also germinate and grow on lung secretions and necrotic lung tissue and produce a mycelium there. However, they cannot invade healthy tissues. The resultant aspergillosis of the lung does not appear to be that important in man, but may cause considerable losses in farm animals and birds. Invasive aspergillosis, invasion and spread from these primary foci, is more serious and common in patients with leukaemia and also where immunosuppressants have been used in lung transplant surgery.

Yeasts, such as *Saccharomyces cerevisiae*, are proving to be suitable vehicles for the biosynthesis of complex, medically important macromolecules, although they do not possess the genetic information to do so. For example, a plasmid containing the gene for the antigen for Hepatitis B has been introduced into yeast cells and they have been used to synthesize the antigen. Such cloned genes have not yet been successfully expressed in bacteria.

Fig. 12.3. Mushroom stones from Guatemala; each is about 300 mm high. (After Heim).

Yeasts, as eukaryotes, may be safer hosts than bacteria for the expression of new genes introduced by genetic engineering.

Certain toadstools have even had a religious significance for man. Among the pre-Hindu pantheon of gods there is one 'Soma' that has a vegetable rather than an animal nature. Reference to it in the Hindu holy books (the Vedas) are obscure, but a strong case has been made for identifying Soma with *Amanita muscaria*. In addition to being rather poisonous, the fly-agaric also contains a potent hallucinogen. Until quite recently certain primitive peoples in Siberia have been known to eat this fungus in the dried state to produce inebriation and erotic intoxication. The resulting visions may have endowed *A. muscaria* with god-like attributes. Again strongly hallucinogenic species of the genus *Psilocybe* have, apparently, long been consumed in religious ceremonies in Mexico. In remote parts of that country the practice may still continue, but in a Christian context. In Guatemala the Maya of 1000–2000 years ago carved fungus-stones, no doubt of religious significance, in which a man or an animal was incorporated in the stipe of the toadstool (Fig. 12.3).

Further reading
Books on fungi

Had reference to the literature been made throughout this book, the text would have been rather congested and a bibliography of many pages would have been needed. It may, however, be of interest to suggest some books on mycology for those concerned with a deeper study of the subject. These are arranged alphabetically under authors. In compiling such a list it is extremely difficult to decide where to stop. Not all the works listed are still in print.

Ainsworth, G. C. (1976) *Introduction to the History of Mycology*, Cambridge University Press.

This splendid and sensitively illustrated book can be read with pleasure by anyone who has developed a strong interest in fungi and is concerned with how their study has developed from classical times to the present.

Alvin, K. L. (1977) *The Observer's Book of Lichens*, Warne, London.

A sound pictorial introduction to the field-study of lichens.

Barron, G. L. (1977) *The Nematode-destroying Fungi*, Canadian Biological Publications, Lancaster, Pennsylvania.

An exciting account of these most intriguing fungi.

Beckett, A., Heath, I. B. and McLaughlin, D. N. (1974) *An Atlas of Fungal Ultrastructure*, Longman, London.

This is a useful pictorial survey of the fine structure of fungi as revealed by the electron-microscope.

Berry, D. J. (1982) *The Biology of Yeast*, Arnold, London.

This booklet describes the structure, physiology, genetics and industrial importance of *Saccharomyces cerevisiae*.

Buller, A. H. R. (1909–34) *Researches on Fungi*, Vols. I–VII, Longmans, Green & Co., London.

These splendid volumes, easy to read and wonderfully illustrated, are available in most scientific libraries. They will reward any student who turns to them. In particular, Vol. III may be recommended for its brilliant accounts of the organization of basidiocarps in *Coprinus*.

Burnett, J. (1976) *Fundamentals of Mycology*, 2nd edn, Arnold, London.

This is a stimulating textbook with a modern approach having its emphasis on the physiology, biochemistry and genetics of fungi.

Dickinson, C. H. and Lucas, J. A. (1982) *Plant Pathology and Plant Pathogens*, 2nd edn, Blackwell Scientific Publications, Oxford.
This is a good introduction to all aspects of plant pathology and includes references to bacteria and viruses.

Garrett, S. D. (1980) *Soil Fungi and Soil Fertility*, 2nd edn, Pergamon, Oxford.
A masterly treatment of the problems of soil mycology.

Gray, W. D. (1959) *The Relation of Fungi to Human Affairs*, Holt, New York.
Deals with all aspects of applied mycology with particular stress on the industrial uses of fungi.

Gregory, P. H. (1973) *The Microbiology of the Atmosphere*, 2nd edn, L. Hill, Aylesbury.
This book discusses in a masterly way the principles governing the spread of micro-organisms, especially fungi, through the air.

Griffin, D. H. (1981) *Fungal Physiology*, John Wiley, New York.
There are several books on fungal physiology but most are now out-of-date. This one gives a well-balanced modern account of the subject.

Hale, M. E. (1983) *The Biology of Lichens*, 3rd edn, Arnold, London.
A good illustrated introduction to the structure, physiology, chemistry and ecology of lichens.

Harley, J. L. and Smith, S. E. (1983) *Mycorrhizal Symbiosis*, Academic Press, London.
There has been much new work on mycorrhiza during the past few years and this book is an authoritative treatment of the subject.

Hudson, H. J. (1986) *Fungal Biology*, Arnold, London.
This gives an account of the role that fungi play in their natural environment.

Isaac, S. (1992) *Fungal–Plant Interactions*, Chapman & Hall, London.
There has been much recent work on fungal/host interactions and this is an authoritative treatment of the subject.

Ingold, C. T. (1971) *Fungal Spores: their Liberation and Dispersal*, Clarendon Press, Oxford.
This gives a detailed account of dispersal in fungi.

Lange, M. and Hora, F. B. (1963) *Collin's Guide to Mushrooms and Toadstools*, Collins, London.
This, with over 600 species figured in colour in 100 plates, is probably still the best of the many good pocket-guides with coloured illustrations now available.

Large, E. C. (1940) *The Advance of the Fungi*, Cape, London.
This reads like a novel but its mycology is consistently sound. It gives a vivid picture of the history of fungi as plant pathogens and can be most strongly recommended.

Manners, J. G. (1982) *Principles of Plant Pathology*, Cambridge University Press.
A modern account of the subject with the chief emphasis on fungi since they are the principal pathogens of plants.

Phillips, R. (1981) *Mushrooms and Other Fungi of Great Britain and Europe*, Pan, London.

This is an essential book for anyone wishing to identify the larger fungi. It is comprehensive and illustrated by colour photographs of the highest quality. Its large format, however, renders it unsuitable for use in the field.

Ramsbottom, J. (1953) *Mushrooms and Toadstools*, Collins, London.
A most readable book illustrated by colour photographs. It gives an engaging account of fungi and their general biology.

Webster, J. (1980) *Introduction to Fungi*, 2nd edn, Cambridge University Press.
This fine textbook, richly illustrated with original figures, deals with morphological, taxonomic and biological aspects of fungi. There is a bibliography of over one thousand titles.